OPERATIONAL RISK MANAGEMENT

OPERATIONAL RISK MANAGEMENT

C. ARIEL PINTO
LUNA MYLENE MAGPILI
RAED M. JARADAT

MP **MOMENTUM** PRESS
ENGINEERING

MOMENTUM PRESS, LLC, NEW YORK

Operational Risk Management

First published by Momentum Press®, LLC
222 East 46th Street, New York, NY 10017
www.momentumpress.net

ISBN-13: 978-1-60650-530-4 (print)
ISBN-13: 978-1-60650-531-1 (e-book)

Momentum Press Industrial and Systems Engineering Collection

Collection ISSN: 2372-3564 (print)
Collection ISSN: 2372-3572 (electronic)

Cover and interior design by Exeter Premedia Services Private Ltd., Chennai, India

10 9 8 7 6 5 4 3 2 1

Printed in the United States of America

ABSTRACT

Businesspersons—including engineers, managers, and technoprenuers—
are trained and drilled to make things happen. Part of their practice is
to guide others on building monuments of success and to make difficult
decisions along the way. However, they will all realize that the decisions
they make eventually determine the chances they take, and that they are
all fraught with uncertainty. This book is developed to give businesspersons the opportunity to learn operational risk management from a systems
perspective and be able to readily put this learning into action, whether in
the classroom or the office, coupled with their experience and respective
discipline.

KEYWORDS

operational risks, risk analysis, risk management, systems approach

CONTENTS

LIST OF FIGURES

LIST OF TABLES

PREFACE

Businesspersons—including engineers, managers, and technoprenuers—are trained and drilled to make things happen. Part of their practice is to guide others on building monuments of success and to make difficult decisions along the way. However, they will all realize that the decisions they make eventually determine the chances they take, and that they are all fraught with uncertainty. This book is developed to give businesspersons the opportunity to learn operational risk management from a systems perspective and be able to readily put this learning into action, whether in the classroom or the office, coupled with their experience and respective discipline. This book is organized around two topics: operational risk management and systems thinking.

Chapter 1 describes operational risk management through fundamental concepts of *accident* (event that is both unintended and undesirable), *hazard* (objects, actions, processes, or condition that may contribute toward the occurrence of an accident), and of course, *risk* (event with undesirable consequences without specific regards to intent). The notion of systems approach is introduced to provide a convenient way to describe, classify, and categorize risk events based on their causes, origins, and consequences.

Chapter 2 provides the generalized framework for operational risk management summarized in these questions: *What should go right? What can go wrong? What are the consequences? What is the likelihood of occurrence? What can be done? What are the alternatives? And what are the effects beyond this particular time?* The notion of the causal chain of events is introduced to where interactions between elements of a system are used to assess the chances of occurrence and the consequences if risk events occur.

Chapter 3 describes several common tools or techniques, namely, Preliminary Hazard Analysis (PHA), Hazard and Operability Analysis (HAZOP), Job Safety Analysis (JSA), Failure Mode and Effects Analysis (FMEA), Fault Tree Analysis (FTA), and Cause and Consequences Analysis (CCA). The principle of As Low As Reasonably Practicable (ALARP), used in managing operational risks, is described to give guidance in applying the tools.

Chapter 4 describes the risk treatment for affecting particular risk events deemed important based on the chances of occurrence, the consequences if the risk events happen, and other factors. The general objectives in risk treatments are (a) reduction in the chances of occurrence of the risk events and (b) reduction in the consequences if the risk events occur. The fundamental steps and corresponding guide questions in developing a plan for risk treatment are described as well.

Chapter 5 describes performance monitoring to control and manage particular operational risk events to within tolerance level and ensure the appropriateness and effectiveness of the risk treatment strategies. To achieve a reliable monitoring process of the operational risk, risk indicators are identified and, information from risk mitigation strategies and risk database are collected.

Chapter 6 describes difficulties for risk managers in coping with the increasing complexities of systems to successfully address complexity. A shift of paradigm with a more holistic perspective is described not only from the technical but also the inherent human–social, organizational–managerial, and political–policy dimensions based on systems thinking and systems theory.

Chapter 7 provides a nonspecific domain system thinking tool that can help risk managers develop systems thinking. The outcome of this tool is a set of systems thinking characteristics to assist in the identification of individuals such as risk managers and members of the risk management team who have the ability to more successfully identify risk and integrate safety in complex systems that require a high level of systems thinking.

ACKNOWLEDGMENTS

The authors gratefully acknowledge the following individuals for their encouragement and support in the writing of this book.

Bill Peterson for providing ideas and guidelines all throughout the development of this book.

Zikai Zhou from the Department of Engineering Management & Systems Engineering at Old Dominion University for his help in gathering important pieces of information that are contained in this book.

Erickson Llaguno from the Department of Industrial Engineering and Operations Research at the University of Philippines for his initial input in the structure of this book.

PART I

FUNDAMENTALS OF OPERATIONAL RISK MANAGEMENT

CHAPTER 1

INTRODUCTION TO OPERATIONAL RISK

1.1 BRIEF HISTORY OF OPERATIONAL RISK

In the early 1990s, the term operational risk was first officially used in the finance industry with the purpose of identifying fraudulent financial reporting. However, by the early 2010s, operational risk, and the management thereof, has acquired a broader description and acceptance. Nowadays, it can be broadly described as having the goal to reduce or mitigate the possibility of something going wrong in any type of operation including, but not limited to, financial operations.[1]

1.2 ACCIDENTS, HAZARDS, AND RISKS

These three terms—accidents, hazards, and risks—are very important, related, but distinct concepts that form the foundation of operational risk management.

Accidents (noun) commonly refer to events that are not intended to happen. Accidental (adjective) refers to the nature of occurrence of events as being unintended. As an example, a fender bender between motor vehicles is an accident related to their operation on public roads because it is generally unintended. In contrast, a fender bender during a demolition derby is not an accident because it may have occurred with intent. Common usage of the term accident may also imply someone's preference for the event—being desirable or not desirable. Fender benders are events that for drivers are accidents that are both unintended and undesirable. Nonetheless, there are instances of fortunate accidents in the common usage of the word, those events that may have not been intended to happen but turned out to be a good thing. Such is the case of the fortunate accidental discovery of penicillin by Alexander Fleming when fungus contaminated his bacterial cultures by accident. Other descriptions of accident are as follows:

- An undesirable and unexpected event, a mishap, an unfortunate chance or event;
- Any unplanned act or event that results in damage to property, material, equipment, or cargo, or personnel injury or death when not the result of enemy action.[2]

For the purpose of this book on operational risk management, the term accident will be used to pertain to events that are unintended and undesirable.[3]

Hazards (noun) usually refer to objects, actions, processes, or conditions that may contribute toward the occurrence of an accident. Hazardous (adjective) refers to the property of objects, actions, processes, or situations that contribute toward an accident. Going back to the example of a motorist, a flat tire (*noun*: a tire that is deflated) is an object that may contribute toward the occurrence of a collision—unintended and undesirable (the accident). Figure 1.1 illustrates the relationship between hazards and an accident. In this illustration, the holes in the slices of cheese that contributed to the accident align and are considered hazards. There are many other holes that do not contribute to the accident and hence are not considered as hazards for this particular accident (but remain hazardous—e.g., potential to cause a different accident).

Risks, or equally *risk events* (noun), refer to future events with undesirable consequences without specific regard to intent, and hence include

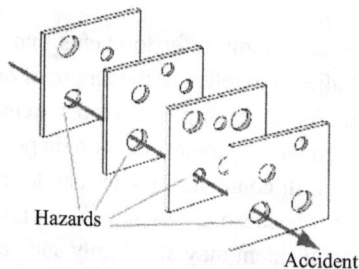

Figure 1.1. Relationship between hazards and an accident.[4]

accidents and non accidents. Riskiness (adjective) is the expression of the magnitude of various properties associated with risks. These properties are usually the likelihood or chance of occurrence, and the consequences if they occur. The following are other definitions of risk:

- The chance, in quantitative terms, of a defined accident occurring;
- Something that might happen and its effect(s) on the achievement of objectives;
- A set of opportunities open to chance.[5]

Fender benders can be considered as risk events because they *may* result in property *damage or worse*. Note the use of the terms *may* and *damage or worse* to emphasize that we are referring to a future event with undesirable consequences.

In the same suit, one may say that tired and distracted drivers have a *higher risk* of being involved in a fender bender. Note the use of the term *higher risk* to describe the higher magnitude of one property of risk events—the chance of occurrence.

Table 1.1 emphasizes the importance of intent and consequence in having a common understanding of the concepts of accidents, hazards, and risks. Events being accidental in turn determine the distinction between what may be construed as hazardous or not. However, risk is more closely related to the consequence rather than intent.

1.3 IMPORTANCE OF A SYSTEMS APPROACH

1.3.1 WHAT IS A SYSTEM?

A system can be described as a collection of interrelated elements contributing toward a common goal. As an example, an automobile can be described as a system made up of a collection of elements (transmission, power, electronics, and other elements) all contributing toward the common goal of transporting passengers and cargo. System may also be the production system that produces automobiles, the company that owns this production system, the supply chain that brings the automobiles to the buyers, and so on. In essence, the scale by which we can view an entity as a system is wide.

Table 1.1. Comparison of accident, hazard, and risk

Accident *noun* Event that is both unintended and undesirable <hurt in an *accident*> **Accidental** *adjective* Being unintended <*accidental* damage>
Hazard *noun* Objects, actions, processes, or condition that may contribute toward the occurrence of an accident <debris on the road is a *hazard* for road accidents> **Hazardous** *adjective* Potential to cause an accident; Actually contributing toward a particular accident; or Causes a particular accident, partly or entirely.
Risk *noun* (also **risk event**) Event with undesirable consequences without specific regards to intent Includes accidents **Risky** *adjective* High magnitude of various properties associated with risk <skydiving is *risky*> <drinking and driving is risky behavior> **Riskier** *adjective* Higher chance of occurrence of a risk event or higher consequence if they occur, or both <skydiving is *riskier* in bad weather>

1.3.2 WHAT ARE THE WAYS TO CHARACTERIZE A SYSTEM?

A system can be characterized by the means of the elements that make them up, their objectives, or the nature of their interrelationships. For example, a modern automobile can be characterized by its automatic transmission, four-wheel drivetrain, electronic ignition, six cylinder engines, antilock brakes, and other elements that make up the particular automobile. This same system can also be characterized by how its elements work together, that is, how the engine's power is relayed by the transmission to the wheels and how the brakes stop the automobile when needed.

1.3.3 WHAT IS A SYSTEMS APPROACH?

A systems approach pertains to the approach of analyzing an entity by recognizing the elements that make it up, their objectives and their interrelationships. This approach is often employed by analysts in various fields or disciplines of applied sciences, for example, engineering, pure sciences, and social sciences.

1.3.4 WHAT IS THE IMPORTANCE OF RECOGNIZING GOALS AND OBJECTIVES?

For the purpose of operational risk management, recognizing goals and objectives of elements as individuals and the system that they form as a whole would lend credibility to an analyst's statement of whether the events are intentional or not and if the consequences are desirable or undesirable. As the previous section emphasized, intent and consequences are important in describing accidents, hazards, and risks, which can be separately viewed as obstacles or impediments to the achievement of goals and objectives.

1.3.5 WHAT IS THE IMPORTANCE OF RECOGNIZING INTERRELATIONSHIPS AMONG ELEMENTS OF A SYSTEM?

Part of effective operational risk management is understanding the sequence of events that eventually lead to accidents. This sequence of events, in many ways, is within the realm of how various elements of the system are interrelated. This recognition of sequence of events is important in any attempt to prevent the occurrence of a risk event or reduce the consequence if it does occur.

1.3.6 WHAT IS THE IMPORTANCE OF RECOGNIZING SYSTEM BOUNDARIES?

Part of operational risk management, particularly the management aspect, implies some degree of control on how elements of a system will behave or relate to each other. Recognizing system boundaries is an acknowledgement of the degree of control one may have over these elements. Elements that, for some reason or another, do not lend themselves to control and are often designated as outside the system boundary. These elements are also termed as external to the system or belonging to the system's environment. However, these external elements are still relevant parts of the systems approach. Hence, the application of a systems approach is to adapt systems thinking by having a thought process that develops an individual's ability to speak and think in a new holistic language. Systems thinking will be discussed in more detail in Chapter 6.

1.4 OPERATIONAL RISK AND OPERATIONAL RISK MANAGEMENT

1.4.1 WHAT IS OPERATION?

In the context of a systems approach, operation can be described as a sequence of processes designed to attain the objectives of the elements or the system as a whole. The operation of an automobile may be described as processes in the following sequence:

1. Fastening the seatbelt
2. Engaging the brake pedal
3. Turning on the ignition
4. Placing the gear in the right setting
5. Disengaging the brake pedal
6. Engaging the gas pedal
... and so on

All these processes are part of operating an automobile to transport passengers and cargo. In a production system, operations may pertain to the sequence of processes—obtaining raw materials, performing mechanical and chemical processes, storage, inspection, movement, and eventual delivery to the users—all designed to attain the goals of the production system.

1.4.2 WHAT IS OPERATIONS MANAGEMENT?

Operations management pertains to the design and control of the operations, usually in the area of production and providing services. As an example, in building a new production line, part of operations management is the design (or procurement) of the machines that will be needed, the identification of needed skills and the hiring of personnel with these skills, the raw materials needed, and a multitude of other items. Once this production line is physically installed, part of operations management is the control of the production line to produce the product that it is designed to produce at the right quantity and quality and at the right time.

1.4.3 WHAT ARE THE OBJECTIVES OF OPERATIONS MANAGEMENT?

The overall objective of operations management on a particular system is to design and control operations to attain the system goals. For a production system, some of these goals are possibly the following:

- Produce the right quantity and quality products.
- Minimize shortages in raw materials, machine resources, and needed personnel.
- Minimize total cost of production.
- Minimize disruption in delivery of finished products.
- Minimize the number of defective products.
- Maximize return (e.g., profit).

For operating an automobile, some of the goals are possibly the following:

- Transport the occupants and cargo from origin to destination.
- Avoid traffic accidents.
- Avoid traffic violations.
- Minimize travel time and cost.

1.4.4 WHAT IS OPERATIONAL RISK MANAGEMENT?

The concept of risk has been described earlier, simply as potential events with undesirable consequences (noun). The following are some of the ways of describing operational risks:

"the risk of loss resulting from inadequate or failed internal processes, people, and systems or from external events"[6]

"a measure of the link between a firm's business activities and the variation in its business results"[7]

This book will subscribe to a modified definition from Basel II—the second of the Basel Accords published in 2004, which are recommendations on banking laws and regulations issued by the Basel Committee on Banking Supervision. Basel II describes operational risk as "the potential undesirable consequences directly or indirectly resulting from fail-

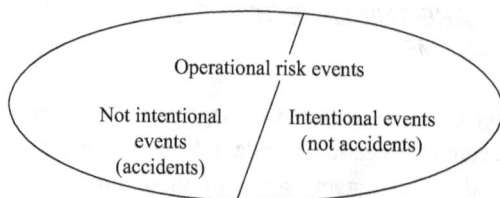

Figure 1.2. Relationship between operational risk events and accidents.

ure of one or more elements of the system of interest."[8] Thus, the term operational risk management may be described as the design and control processes that will affect operational risks.

It should be noted that the word *failure* is used, implying that operational risk events are not intended results of system operation. Nonetheless, all operational risk events cannot be deemed as accidents because in modern complex systems, intent behind undesirable events may only be established after lengthy and careful investigation.

The relationship between possible operational risk events and accidents is illustrated in Figure 1.2. Consider as an example the inaccurate transfer of money from one bank account into another—a particular operational risk event in financial institutions. Consider further that after a thorough investigation, it has been established that this particular instance was not intentional and may have been a result of a software bug or honest human mistake. Only then can this instance be considered as an accident. However, if it has been established as a fact that this particular instance was an intentional result of human actions, then it is considered as fraud and is not an accident. Either way, whether an accident or not, this is still an operational risk event. Figure 1.2 illustrates that operational risk management is concerned with all undesirable events that may happen in a system regardless of whether they are intentional or not. This includes both accidents and non accidents.

1.5 CLASSIFICATION OF OPERATIONAL RISK EVENTS

Operational risk events are often classified in some manner to help analysts and managers to better address the risks. The manner of categorization depends on the field of application and industry. One way to categorize operational risks is by identifying distinct elements of the system that can be construed as the primary cause of the risk event. Table 1.2 enumerates, describes, and provides examples of risk events classified based on their primary cause. Operational risk can be classified by the usual classification of elements that make up a system and its environment: people, process, information, materials, machine, and external events.

Another way of categorizing operational risk events is based on the origin of the events. Table 1.3 describes some of these classifications.

Table 1.2. Classification of operational risk based on primary cause[9]

Classification	Description	Examples
Human	Risk events that are primarily caused by humans, both intentional and not	Fraud; breaches of employment law; unauthorized activities; loss or lack of key personnel; inadequate training; inadequate supervision; fatigue; lack of action
Process	Risk events that are primarily caused by activities performed by elements of the system	Designed or authorized process not fit for purpose; errors caused by models
Information	Risk events resulting from inaccurate, false, or untimely information	Late information; corrupted data; inaccurate information, mismatched unit of measure
Materials	Risk events resulting from defective or inappropriate materials	Concrete that are too soft; metals that corrode too easily; plastics that are too toxic
Machines	Risk events caused by malfunctioning machines and equipment	Machine breakdown; imprecise machine performance
External events	Risk events caused by elements external to the system	Cyber hacking; failure in outsourcing; extreme weather; unfair regulations; political issues; unfair competition; fragmented market

Table 1.3. Classification of operational risk based on origin[10]

Classification	Description	Examples
Organizational	Risk events originating from humans and their roles in the organization	Noncompliance to regulation; under-skilled technicians
Technical	Risk events originating from nonhuman aspects of the system	Late information; machine break-down
Social	Risk events originating from humans whose roles may be outside the organization	Crime; unfair competition
Political	Risk events originating from policies and regulation imposed by government or other authorities	Unfair regulations
Environmental	Risk events originating from the natural environment	Extreme weather

Table 1.4. Classification of operational risk based on consequence

Classification	Description	Examples
Safety	Risk events resulting in injury, death, or failure in human health	Machine failure resulting in injury to the operator
Financial	Risk events resulting in financial losses	Unfair regulations resulting in lost revenue
Legal	Risk events resulting in legal suits	Noncompliance to regulation resulting in suit from the government
Security	Risk events resulting in deterioration of protection of valuable assets	Cyber hacking resulting in theft of confidential information

Risk events may also be classified based on its consequence or affected elements of the system. Table 1.4 describes some of these classifications.

It is noticeable from Tables 1.2 to 1.4 that a particular risk event may be classified in various ways depending on the classification strategy. As an example, a machine failure resulting in injury or death may be classified as a machine-related risk (because it is primarily caused by the machine), a technical risk (because a machine is included under technical management), and safety risk (because the consequence may be injury or death). These apparent cross-classifications may be of benefit for the analyst and manager because it provides a richer description of the particular operational risk event—from its primary cause, the management area that it may commonly fall into within the organization, and the consequence. This rich description of the particular operational risk will help toward the effective treatment of risk events, as will be discussed in Chapter 4.

1.6 DIFFERENCE BETWEEN OPERATIONAL RISK AND OTHER TYPES OF RISKS

This book describes operational risks as the potential undesirable consequences directly or indirectly resulting from the failure of one or more elements of the system of interest. However, there is no attempt in this book to imply that this is the only description of this term because there are other reference materials and practitioners that subscribe to a narrower use of this term. Consider terms such as liquidity risks, market and product risks, and group risks. It can be argued that these risks, pronounced in the operation of financial institutions, overlap but are not exactly the same as operational risks.[11] However, it can also be argued that distinction may be resolved by careful definition of the boundary of the particular system. Furthermore, such boundaries may be artificial because further analysis may show the fact that many of the consequences ascribed to other classes of risk are coupled with operational risk events.

In manufacturing, where several systems work together to create the supply or value chain, a risk event occurring in one system may have a far reaching effect extending to other systems. In the global nature of automotive manufacture, a political risk in one geographical region where a supplier of upholstery fabric is located can eventually affect the operation of final assembly lines for cars across the globe. The fact is that operational risk crosses political, geographical, managerial, organizational, and corporate boundaries.

The following list shows some examples of real cases of operational risk events from various industries and geographical region.

- Sumitomo (1996): The disclosure of the $2.6 billion copper-trading-related loss blamed on an employee through unauthorized trades.[12]
- Barings Bank (1995): The bank collapsed in 1995 after one of the bank's employees lost $1.3 billion due to unauthorized speculative investing.

- Chernobyl nuclear reactor disaster (1986): The Chernobyl disaster was a catastrophic nuclear accident that occurred on 26 April 1986 at the Chernobyl Nuclear Power Plant in Ukraine (then officially the Ukrainian SSR), which was under the direct jurisdiction of the central authorities of the Soviet Union.[13]
- Collapse of Maxwell Communications (1991): Maxwell Communications Corporation was a leading British media business where hundreds of millions of pounds were stolen from employees' pension funds.
- Foot-and-mouth disease crisis in the United Kingdom (2001): The epidemic of foot-and-mouth disease in the United Kingdom in the spring and summer of 2001 was caused by the "Type O pan Asia" strain of the disease. This episode resulted in more than 2,000 cases of the diseases in farms throughout the British countryside.
- Northeast United States power failure (2003): The Northeast blackout of 2003 was a widespread power outage that occurred throughout parts of the Northeastern and Midwestern United States and the Canadian province of Ontario on Thursday, August 14, 2003, just before 4:10 p.m. EDT(UTC−04).
- Lac-Mégantic train accident (2013): Occurred in the town of Lac-Mégantic, located in the Eastern Townships of the Canadian province of Quebec, on July 6, 2013, when an unattended 74-car freight train carrying Bakken formation crude oil ran away and derailed, resulting in the fire and explosion. Forty-two people were confirmed dead.[14]

1.7 SUMMARY

Operational risk management is founded on the concepts of *accident* (event that is both unintended and undesirable), *hazard* (objects, actions, processes, or condition that may contribute towards the occurrence of an accident), and of course, *risk* (event with undesirable consequences without specific regards to intent). A systems approach provides a convenient way to describe, classify, and categorize risk events based on their causes, origins, and consequences. Even though there are other types of risks besides operational risks, such distinction can usually be clarified by looking at them from the systems perspective.

CHAPTER 2

THE RISK MANAGEMENT PROCESS

There are many risk management processes used in various industries, disciplines, and professions. These risk management processes differ primarily in the details of how the risk management process is undertaken. These variations mainly stem from differences in the specific systems, conditions, and environments that are unique to a particular industry or discipline. For example, the risk management process for building highways would be different from the risk management process for processing hazardous chemicals because the things that can go wrong (i.e., risks) for these industries are distinct in terms of their specific hazards, causes, and consequences. Hence, as part of the risk management process, it is important that the system be well defined.

It is necessary to describe the system of interest as well as its boundaries to focus the efforts of the risk management process. Risk management can be and should be applied to the total system as a whole. However, when necessary, risk studies and analysis can focus on certain subsystems, processes, or components of concern. The scope can range from large complex systems such as production facilities to small subsystems or components of a product.

The boundaries of the system can be established by first defining the purpose or objective of the system and what elements support the achievement of this purpose. It helps to then identify which elements are within your control or influence and which are not.

2.1 GENERAL GUIDING QUESTIONS IN RISK MANAGEMENT

Traditional risk management processes try to answer five guiding questions: (a) *What can go wrong?* (b) *What are the causes and consequences?*

Table 2.1. 1 + 6 General guiding questions in risk management

0th What should go right?
1st What can go wrong?
2nd What are the causes and consequences?
3rd What is the likelihood of occurrence?
4rd What can be done to detect, control, and manage them?
5th What are the alternatives?
6th What are the effects beyond this particular time?

(c) *What are the likelihood or chances of occurrence?* (d) *What can be done to detect, control, and manage them?* and (e) *What are the alternatives?* These five questions correspond to the three stages in risk management[1] referred to as risk identification, assessment, and mitigation, respectively.

A more progressive risk management process based on the systems approach includes an extension of these questions,[2] in particular, *What should go right?* a precursory question to *What can go wrong?* Asking what needs to go right is common in many engineering and management endeavors and is a basic aspect of any design process, hence labeled as *0th guiding question* to emphasize that this question is not exclusive to risk management. Although it may sound trivial, the underlying principle in this first step is that to know what can go wrong, one must first know what should go right.

Another extension to the traditional questions is an assessment question regarding the future—*What are the effects beyond this particular time?* This reasons that whatever has been recommended and implemented to address risks should be reviewed for relevance and effectiveness not only for conditions that exist now but also for conditions that may occur in the future. It also makes an assessment of the impacts (intended or unintended) of the recommendation beyond the system of interest, such as its effects on the system's environments or other systems. Table 2.1 lists the extended questions of the risk management process.

2.1.1 0TH WHAT SHOULD GO RIGHT?

Once the system of interest is defined, its primary goals and related constraints need to be identified. Consider the examples provided in Chapter 1 for a production system in which the following are some of the possible goals:

- Produce the right quantity and quality products.
- Minimize shortages in raw materials, machine resources, and needed personnel.
- Minimize total cost of production.
- Minimize disruption in delivery of finished products.
- Minimize the number of defective products.
- Maximize return on invested resources (e.g., profit).

2.1.2 1ST WHAT CAN GO WRONG?

Once the system of interest is defined, identifying the hazards, risks, and accidents that may happen in the operation of a system is the next step in managing risk. It will not be a stretch of analysis to deduce some undesirable events based on the objectives, for example:

- Failure to produce the right quantity and quality products.
- Excessive shortages of raw materials, machine resources, and needed personnel.
- Unacceptable total cost of production.

Negative scenario identification (also known as antigoal, antiobjective, or negative scenario analyses) is a common strategy that basically conceives different ways things can go wrong in a system based on what are known as desired events. It helps to imagine departures or deviations from the ideal. Still, identification of risk can be a daunting and difficult task. One can begin by exploring and searching for risks, as it relates to workers, the materials used or produced, equipment and machines employed, process and operating conditions, information, and the external environment such as regulatory requirements. In Chapter 1, Tables 1.2 to 1.4 provide some classification of operational risk events based on primary causes, origins, and consequences, which can all serve as a starting point in identifying what can go wrong.

There are also several ways to find information on a particular risk, hazard, or accident described as follows.

2.1.2.1 Using Historical Information

The simplest way to identify risk that may happen in the future is by looking back at the past. This is particularly applicable for systems that are currently operating or systems that have operated or existed in some form in the past. Histories of risk events (both accidents and nonaccidents) that have happened are usually documented in various forms: written, electronic, or tacit. Companies that are diligent in their risk management practices, usually keep detailed information surrounding these events in the form of a risk register, a risk log, or hazard log.

Tacit information of past risk events on the other hand are primarily "stored" in the form of anecdotes from personnel who may have been present during their occurrence or who may have learned of the events from other colleagues. To transform such information to nontacit, many corporations use methods such as interviews and group discussions. These methods can be captured and converted into a more stable and accessible format in hard copy and electronic forms such as minutes of meetings, audio recordings, and video recordings. Tacit information whenever collected should be included in risk registers.

As an example, consider the event of a power supply interruption in a chemical processing plant. The following information is commonly found in risk registers related to this risk event:

- Brief description of risk event—for example, power supply interruption during time of operation of chemical processing;
- Known causes—for example, power outage at local electrical grid supply, accidental switch off of power switch into the facility, localized natural events such as lightning strikes, localized accidents such as fire, and others;
- Known consequences—for example, interruption in the affected facility and all ongoing chemical processes, downtime, production delays, product defects, and others;
- Risk management strategies in place—for example, uninterrupted power supply by backup generator.

In addition, risk registers may also include details such as observed frequency of occurrence, secondary consequences, known hazards, and effectiveness of management strategies. Particularly, attention is given to those risks with significant undesirable consequences such as considerable financial loss, injury, and death. Human or worker-related risks can be found in injury and illness records, accident and incident accounts, workers compensation claims, and regulatory agency reports.

2.1.2.2 Using Comparative Analysis

There are instances in which identifying risk from historical information may not be sufficient or even possible, as is the case with a system not yet built or still being developed. In these cases, there are no past records of risk events. Another instance is when a system is undergoing significant change. For example, an existing production system that is adding a new product line may have significant changes in personnel, equipment, and process, or the environment such as prevailing economic conditions or newly enacted laws and regulations may change. In these cases, past records on risk events may not be relevant because of changes within the system and its environment. One method to identify new risk events uses comparative analysis. Observing other *similar* systems and risk events that have occurred in those systems becomes useful information. Observing similar systems includes looking at historical information such as risk registries and hazard logs of those systems that are comparable.

2.1.2.3 Using Predefined List of Risk Events

Data banks of risk registers or hazard logs provide a wealth of information on risk events. Risk registers are again particularly useful in extracting risk events that may not have occurred in a given system but have been observed in other similar systems. This type of list is commonly published by federal agencies as well as professional and industry groups based on a compilation of risk registers from systems under their purview. These come in the form of lists, checklists, records, databases, standards, and codes of practice. For example, the construction industry has pre-existing hazard lists compiled by the Occupational Safety and Health Administration (OSHA). Consider as an example a list of risk events that applies to construction workers. A predefined hazard list includes the following:

- Falling
- Sunburn
- Exhaustion
- Unacceptable loud noise
- Physical injuries
- Dust and particulate inhalation

Manuals, handbooks, and guides also often contain predetermined list of known risks. Material safety data sheets, equipment manuals, equipment manufacturers' technical representatives, existing health and safety plans, and handbooks can help identify common risks.

2.1.2.4 Team-Based Elicitation

Brainstorming is particularly useful when hazards have to be conceived (abstractly imagined) and predicted (foreseen). A multidisciplinary cross-functional team with specific knowledge on the system can help identify many potential risks. This method capitalizes on the wide range of knowledge and experience of team members, especially tacit knowledge. Representatives in the organizational functions related to the system of interest should be included, starting with the operators, supervisors, design and quality engineers, maintenance personnel, and all other workers who directly interact with the system of interest. Occupational and safety experts, health specialists, and other personnel experienced with industry-specific risks are also good candidates to be included as team members. Collective expertise and experience is important to ensure that as many potential hazards as possible are identified. A meeting or workshop format may be helpful in gathering data quickly. Questionnaires, surveys, and interviews facilitate individualized inputs and can gather more information from a wide variety of sources. Even in the few instances where the risk management process is an individual endeavor, efforts must be made to gather inputs from as many sources as possible.

2.1.2.5 Modeling

Models are also used to identify potential risk events. Models include analytical, mathematical, physical, mental, and other types, which may be used to represent a new system, the changes within the existing system, and the system's external environment. Many models of these types often include computer simulations. Consider as an example the prototype modeling of cars and airplanes to identify which components will most likely fail after extensive use under various scenarios.

2.1.3 2ND WHAT ARE THE CAUSES AND CONSEQUENCES?

Once risk events are identified, the next phase is to describe these events for the purpose of extending understanding and knowledge about the event. This involves establishing causality, identifying *root causes* and their *likelihood*, as well as characterizing *consequences and impact*. This helps in developing more appropriate and effective decisions or actions related to the management of risk. Invoking the notion of the systems

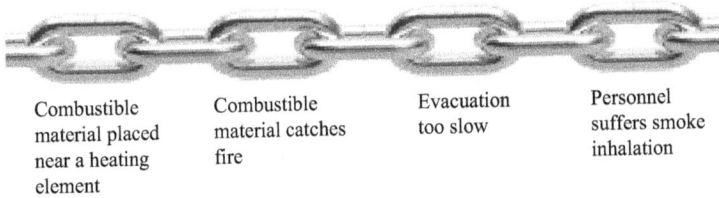

| Combustible material placed near a heating element | Combustible material catches fire | Evacuation too slow | Personnel suffers smoke inhalation |

Figure 2.1. A causal chain of events showing interaction between elements of a system, mechanics of fire, and the risk event of a fire in an industrial facility.

approach has proven to be effective toward describing a particular risk. Establishing what causes accidents, risk events, or any type of events in general can be termed as establishing the *causal chain of events*. The use of the notion of a *chain* is a metaphor for how events form distinct links that together form a continuous span. As an example, Figure 2.1 illustrates how the knowledge of interaction between elements of a system and the mechanics of fire help in establishing a causal chain of events. The role of a causal chain of events in managing risks will be more apparent in the details of many tools and techniques used in analyzing risks—some of which will be discussed in Chapter 3.

Based on the same notion of causal chain events as described earlier, the direct and indirect consequences of a particular risk event are identified. The distinction between direct and indirect consequences is usually based on temporal separation, system boundary, and direct causal attribution. From a systems perspective, consequences of some risk events may propagate beyond the system of interests and hence be considered as indirect consequences. As an example, a production system that fails to produce the right quality product (contrary to its objective) may be directly affected in terms of reduced profit. Nonetheless, failure to produce the right quality product may also affect the morale of the production personnel, indirectly due to stagnant salary. Consequences are often predicted and estimated by looking at how other events may occur (unintended consequences) as a result of the occurrence of the risk event and their impact on the various elements of the system and the achievement of the system's overall objective.

Often the severity of consequences is evaluated through a rating scale such as that shown in Table 2.2. In the table, effects are rated from 1 to 5, where 1 is negligible and 5 is catastrophic. There may be risks that have more than one consequence. In this case, it is typical that focus is placed on the consequence with the highest severity rating. This avoids the com-

Table 2.2. Sample rating scale to estimate consequences of risk events[3]

Severity class	Example severity scale Description	Rating
Catastrophic	Failure results in major injury or death	5
Significant	Failure results in minor injury	4
Major	Failure results in medium to high level of exposure but does not cause injury	3
Minor	Failure results in low-level exposure but does not cause injury	2
Negligible	Failure results in negligible exposure	1

plication of dealing with too many effects with low ratings, enabling the focus to be placed more on the highest rated ones first.

Ultimately, analysis of the causal chain of events leads to the identification of root causes. The knowledge of various elements of a system and their interactions in time may explain how a risk event can eventually occur. The interaction among various elements, for example, how humans use a machine in an industrial process, may be used to establish the causal chain of events. The key is to identify root causes to be addressed. Many cause analysis tools such as root cause analysis, fish bone diagram, Pareto charts, and scatter diagrams can be useful. Other tools such as Fault Tree Analysis (FTA) and Cause-Consequences Analysis (CCA) (discussed in Chapter 3) can also be used for identifying root causes of more complex systems. These types of analysis coupled with knowledge from the team can help determine the root causes of identified risks.

2.1.4 3RD WHAT ARE THE CHANCES OF OCCURRENCE?

Sequences of events that lead to a particular risk event, that is, the causal chain of events, need to be described in terms of their respective chances of occurrence. The frequency or chance of occurrence of a risk event pertains to the quantitative or qualitative description of how often or how soon a particular risk event may occur. This is often derived from historical information or records of the risk event. This can also be derived from team-based elicitation. Complex probabilities can be calculated using tools such as FTA, CCA, or Event Tree Analysis (ETA) (discussed in Chapter 3).

Table 2.3. Sample rating scale to estimate likelihood of risk events[4]

Example likelihood scale		
Likelihood of cause	Criteria: occurrence of cause	Rating
Very likely	Once per month or more often	5
Likely	Once per year	4
Moderately possible	Once per 10 years	3
Unlikely	Once per 100 years	2
Very unlikely	Once per 1000 years	1

Likelihoods are also evaluated through a rating scale such as that shown in Table 2.3. In the table, likelihoods are rated from 1 to 5, where 1 is extremely unlikely and 5 is frequent.

A well-rounded team proves valuable with ratings, as they can be highly subjective and should be provided by people who have direct knowledge and experience with the system and the risk being evaluated.

2.1.5   4TH WHAT CAN BE DONE TO DETECT, CONTROL, AND MANAGE THEM?

Ranking and scoring is conducted to evaluate criticality and determine relative importance. What may be critical is contextual. However, common critical risks are those whose consequences are related to health and safety, compliance to regulatory requirements, or those that affect core business practices. Criticality may be assessed using a risk matrix similar to that shown in Table 2.4. The risk matrix in Table 2.4 highlights risk events with high severity ratings such as those risks that fall under the *catastrophic* category of consequences or risks that fall under the *very likely* category of likelihood of occurrence. However, particular attention should be given to those risks in which consequences are catastrophic *and* the likelihood of occurring is very likely or eminent. In Table 2.4, these are the risk events that fall in the darker boxes. Risk events that fall in the darkest boxes should be addressed immediately. Risk matrix tables are useful for categorizing and prioritizing identified risks.

Moreover, an aggregated approach can be used to determine criticality. A risk score or risk priority number can be derived by combining the ratings on the severity of consequences and likelihood. One of the most common methods is multiplying the severity rating and likelihood rating. For example, a risk with catastrophic consequences (rated as 5 in

Table 2.4. Sample risk matrix

Likelihood of occurrence	Consequence				
	Negligible	Minor	Major	Significant	Catastrophic
Very likely					
Likely					High
Moderately possible	Low	Low medium	Medium	Medium high	
Unlikely					
Very unlikely					

Table 2.2) and a very likelihood of occurrence (rated as 5 in Table 2.3) will have a risk score of 25 (5 × 5). Many times current capabilities to address those risks are also incorporated in the score. This method is shown in Chapter 3's section on FMEA: Failure Modes and Effects Analysis. The risk score can provide guidance for ranking potential risks in the order they should be addressed. The purpose is to concentrate resources, time, and effort to the risks that are most significant.

After the more significant (high priority risk) events are identified, the strategies to treat these risk events are generated based on the information known from its causal chain of events. This stage of the risk management process is also known as risk treatment and is discussed in more detail in Chapter 4. However, it will suffice to think of the following guiding questions when trying to detect, control, and manage risk events.

- *Which segments in the causal chain of events can be detected?* Certain segments in the causal chain of events may be more amenable to detection than other segments based on the mechanics of the phenomena, available technology for detections, and other factors (e.g., ethical and legal factors). Also, how early the start of a risk event can be detected gives you time to address it. Early detection is desired.
- *Which causal chain of events and target risk events can be controlled and how can they be controlled?* In the same way that segments in the causal chain of events may be more amenable to detection, some of these segments may be more amenable to control than others.
- *If the risk event happens, how can the system recover quickly?* Important functions in the operation of the system are identified for the purpose of analyzing what can be done to prevent or reduce the consequences.

2.1.6   5TH WHAT ARE THE ALTERNATIVES?

Which risk treatment strategies will work well together, given the causal chain of events? Risk treatment strategies are not mutually exclusive, and effective action plans are usually made up of combination of strategies, albeit in various degrees. Consider the causal chain of events illustrated in Figure 2.1 of a system and mechanics of fire. Based on common experience, we can identify particular risk strategies to address the various segments in this chain, for example, regular housekeeping to prevent the accumulation of combustible materials near a heating element, installing and maintaining smoke detectors near equipment with heating elements, having an evacuation plan and conducting regular drills, training to prevent smoke inhalation, and others. However, all these strategies can be implemented in various degrees and combinations; the appropriateness may depend on the particulars of the specific system.

In general, risk treatment strategies are identified for reducing chances of occurrence, for reducing consequences if they do occur, or both. *Detection* and *control* are the typical strategies to reduce the chances of occurrence and are often applied in anticipation of a risk event, while *recovery* plans address the reduction of consequences after risk events have occurred. Specific details and examples of these risk treatment strategies are described in Chapter 4.

Regardless of the combinations of risk strategies, the usual criteria in choosing alternatives are

- Early detection
- Minimize direct and indirect consequences
- Faster recovery
- Impede accrual of consequences.

Standards and codes from federal agencies and industry databases that prescribe or suggest strategies and solutions to common hazards and risks may already be available and should be explored first to stimulate discussions and brainstorming. Even so, the comprehensiveness and viability of a holistic strategy benefits most from the team-based approach. Combined experiences and expertise allow for a more thorough analysis of risks and more effective proposals to address them. Further research may need to be conducted to ensure adequate understanding and response to the identified risks. In some cases, one may need to consult specialists to ensure appropriate expertise in dealing with particularly dangerous and complex hazards.

2.1.7 6TH WHAT ARE THE EFFECTS BEYOND THIS PARTICULAR TIME?

From a system's perspective, it is important to evaluate the effects of the risk treatment alternatives to other elements of the system. Risk treatment alternatives may be analyzed according to their effects to functionalities of elements of the system, the manner by which they alter interaction among elements, and their potential to affect future decisions. This is also the point where the acceptable level of risk is determined by comparing the costs and benefits of each mitigation alternative. The concept of As Low as Reasonably Practicable, a fundamental approach that sets the risk to the tolerable, reasonable, and practical level, will be further discussed in Chapter 3.

There is also the notion of residual and emerging risks, which are manifestations of the fact that no risk events can be totally eliminated and that new ones may emerge in the process of treating others. A more detailed discussion is provided in Chapter 4.

2.2 SUMMARY

The generalized framework for operational risk management can be summarized in six steps, with one precursory step (hence, the label 1 + 6 guiding questions in Table 2.1). These questions extend the more commonly known risk analysis, assessment, and management. The questions are as follows: *What should go right? What can go wrong? What are the causes and consequences? What is the likelihood of occurrence? What can be done to detect, control, and manage them? What are the alternatives? And what are the effects beyond this particular time?*

The notion of the causal chain of events was introduced to show interactions between elements of a system and how this can be used to assess the chances of occurrence and the consequences if the risk events occur. The causal chain of events can also be used to generate ways of how initiating events can be detected before a risk event occurs, as well as what can be done to control their occurrence and consequences.

PART II

MANAGING OPERATIONAL RISKS

CHAPTER 3

TOOLS AND TECHNIQUES

This chapter describes several common techniques and a principle used in managing operational risks. The techniques are Preliminary Hazard Analysis (PHA), Hazard and Operability Analysis (HAZOP), Job Safety Analysis (JSA), Failure Mode and Effects Analysis (FMEA), Fault Tree Analysis (FTA), and Cause and Consequences Analysis (CCA). The principle of As Low As Reasonably Practicable (ALARP) is described to give guidance in applying the risk management process. Their basic descriptions are summarized in Tables 3.1a and 3.1b.

3.1 PRELIMINARY HAZARD ANALYSIS

A PHA is a semiquantitative approach used to develop an initial listing of potential hazards, hazardous situations, and hazardous events. The emphasis in a PHA is the breadth of identifying all possible and conceivable hazards. It is thus a broad appraisal of risk events and is typically done early in a product or system development phase or during a project's very early stages of conceptualization. Variants of PHA are Preliminary Risk Assessment (PRA), Rapid Risk Ranking, or Hazard Identification (HAZID).[1]

3.1.1 WHERE AND WHEN TO USE PHA

A PHA is useful as a preliminary study to determine risk requirements in the early stages of the design process, especially safety-related requirements. Hence, it is widely used to support the identification and development of system or product specifications. It can also be used as an initial step of a detailed risk analysis of an existing project, product, or system, that is, a precursor to further detailed studies. It can also serve

Table 3.1a. Summary descriptions of six common tools

PHA
• Initial assessment of the hazards and their corresponding accidents • Identifies hazards, hazardous situations, and events for a given activity, facility, or system

HAZOP
• Structured and systematic technique for system examination and risk management • Identifies and evaluates problems that may represent risks to personnel, equipment, and operation

JSA
• Emphasizes job safety by analyzing workplace hazards • Identifies hazards associated with each step of any job or task that has the potential to cause serious injury

FMEA
• Design tool for the systematic analysis of component failures and their effects on system operations • Identifies potential design and process failures before they occur and proposes changes to design or operational procedures

FTA
• Uses deductive top-down modeling technique to analyze how an unwanted event or a failure can occur • Identifies linkages between segments in a causal chain of events

CCA
• Combines two different types of tree structures for analyzing consequence chains • Shows the way in which various factors may combine to cause a hazardous event along with event trees to show the various possible outcomes

Table 3.1b. A key principle in operational risk management

ALARP
• Fundamental approach that sets the risk to the tolerable, reasonable, and practical level • Provides guidance for when further reducing the likelihood and consequences of risk events may be disproportionate to the time, cost, and physical difficulty of implementing risk treatments

as a complete risk analysis of a rather simple system or where conditions prevent a more extensive technique from being used.[2]

A PHA can be used to initiate the formation of a hazard log. A hazard log, also known as risk register or risk log, tracks information about a hazard from its initial identification to its proposed control measures through the implementation of interventions to address the hazard. A hazard log facilitates the continuous monitoring of a hazard to make sure that it remains under control or is eliminated.

3.1.2 PHA FORMAT

The tabular form or spreadsheet is the most widely used format for PHA. An example of a PHA worksheet for a simple system, an elevator door, is shown in Figure 3.1. The columns of the PHA table generally follow the key questions of the risk management process as discussed in Chapter 2: *What can go wrong? What are the causes and consequences? What are the likelihood or chances of occurrence? And what can be done to detect, control, and manage them?*

These questions provide a good guide to follow when completing a worksheet. The following are the descriptions of columns that correspond to the PHA worksheet shown in Figure 3.1.

- *What can go wrong?*
  - o Column 1: Hazards that may lead to an accident
- *What are the causes and consequences? What are the likelihood or chances of occurrence?*
  - o Column 2: Accidents that can occur from the hazards
  - o Column 3: The potential causes that can bring about each accident
  - o Column 4: The likelihood that a particular cause triggers the accident
  - o Column 5: The severity of the consequences if the accident occurs
  - o Column 6: The risk score based on the likelihood information entered in Column 4 and the severity information entered in Column 5
- *What can be done to detect, control, and manage the hazards?*
  - o Column 7: Recommended preventive and control measures to address the hazards, the causes, and their consequences

System: elevator door

#	(1) Hazards	(2) Accidents	(3) Potential causes	(4) Likelihood (L) Rating scale	(5) Severity (S) Rating scale	(6) Risk score	(7) Possible controls, preventive actions, mitigation
			Description				**Analyst date**
	What are the possible hazards	*What could be the harm of the hazard*	*How might the accident occur*	*What are the likelihood or chances of occurrence*	*How significant is the harm*	$L \times S$ *Ranking*	*What can be done*
1	Elevator door closing on obstruction	Body or body part trapped or pinned by the door which can result in bodily injury, hospitalization, or death	Obstruction is not detected	4	5	20	Door should be able to check pathway for obstruction • Install sensors • Implement regular testing and maintenance of sensors • Implement emergency plan and protocols
			Door mechanism is not working, that is, mechanism does not stop door from closing	2	5	10	An alternative mechanism to bypass standard controls should be available • Install emergency override switch • Implement system testing and maintenance • Implement emergency plan and protocols

Figure 3.1. PHA worksheet of an elevator door.

3.1.3 GENERALIZED STEPS FOR PHA

1. Define the system and scope of study.
2. Identify possible hazards.
3. Describe the accidents that can be caused by the hazards, as well as their consequences.
4. Identify potential causes of how might the accidents identified in step 3 can occur.
5. Evaluate each cause according to the likelihood that it can occur and the severity of consequences.
6. Rank hazards in terms of step 5.
7. Determine recommended preventive and control measures to address the hazards, the causes, and their consequences

Step 1 defines the system and clearly distinguishes the scope of the PHA. As an example, consider an elevator door as a system of interest. For the succeeding steps, refer to the PHA worksheet in Figure 3.1.

Step 2 identifies possible hazards. Recall from Chapter 1, a hazard is defined to refer to objects, actions, processes, or conditions that may contribute toward the occurrence of an accident. For an elevator door, a possible hazard is the door closing on an obstruction. This information is entered in Column 1 of the PHA worksheet.

Step 3 describes the potential accident caused by the hazard. What could be the harm of the door closing on an obstruction? If the obstruction was a body or body part, then getting pinned or trapped is a potential accident that may result in bodily injury, dismemberment, or even death. This information is entered in Column 2 of the PHA worksheet.

Step 4 examines the possible causes of how the accident can occur. For the elevator door closing on an obstruction, two potential causes can be identified: (1) the obstruction is not detected and (2) the door mechanism is not working. This information is entered in Column 3 of the PHA worksheet.

Step 5 evaluates the likelihood and severity for each potential cause. A rating scale such as Tables 2.2 and 2.3 in Chapter 2 is typically adopted. For the elevator door PHA, suppose the number of accidents occurring from the obstruction not being detected is on average one per year, then the likelihood is rated at 4 (see Table 2.3). The absence of sensors on the elevator door accounts for this high likelihood of accidents. For the door mechanism not working, the number of accidents occurs only once per 100 years due to the high quality of the door mechanisms. Using Table 2.3, its likelihood is rated at 2. This information is entered in Column 4 of the PHA worksheet.

The severity of the consequences is also rated according to how serious the impacts are as described in step 3. For the elevator door PHA, bodily injury and death are deemed catastrophic and thus given the highest rating of 5 (see Table 2.2). This information is entered in Column 5 of the PHA worksheet.

Step 6 facilitates the ranking of the hazards and potential causes by determining an overall risk score for each entry. For the elevator door PHA, this is done by multiplying likelihood and severity ($L \times S$). This information is entered in Column 6 of the PHA worksheet. The risk of an accident from the obstruction not being detected is more significant than the risk from the door mechanism not working, with a risk score of 20 and 10, respectively. Thus, addressing the detection of an obstruction should be prioritized.

Step 7 develops recommended measures to prevent, control, or mitigate the hazards and its consequences as it relates to each potential cause. For the elevator door example, proposed measures such as the installment of sensors to detect obstructions is entered in Column 7 of the PHA worksheet.

3.2 HAZARD AND OPERABILITY ANALYSIS

HAZOP is a top-down *qualitative* approach, as opposed to semiquantitative approach such as PHA. It is also used to determine potential hazards in a process, operation, or system. In particular, it systematically identifies possible *deviations* from the design and operating intent and determines how these deviations may eventually become hazards. The term hazard in HAZOP pertains to these deviations and their causes and collectively may be viewed as part of the causal chain events that can lead to an accident.[3] The term operability in HAZOP pertains to the focus on hazards resulting from operations or processes that are performed outside the range of the intended design (hence, again the emphasis is on deviation).

Unlike PHA, HAZOP's qualitative nature does not attempt to quantify or rate hazards. Likelihood and severity ratings are omitted. Thus, it does not typically prioritize risk. Moreover, HAZOP documents existing preventive and control measures that are in place. Recommendations focus on additional measures that are needed.

3.2.1 WHERE AND WHEN TO USE HAZOP

Similar to JSA discussed in a latter section, HAZOP was originally intended for analyzing accidents that affect human health and safety. It has since been adopted as a general-purpose risk management tool. It is applicable to various types of systems, especially those with highly monitored performance and detailed design requirements found in the chemical, petroleum, food industries, and the like. Similar to PHA, it is applicable during the design phase of a project, product, or system development to determine specifications. It is also often applied to existing systems to improve safety and minimize operational risk. It is, in fact, useful for periodic review throughout the life of a system to accommodate changes within the system and its environment. It can also be used as supporting documentation for hazard logs.

3.2.2 HAZOP FORMAT

HAZOP typically relies on the use of system diagrams such as Process Flow Diagrams (PFD) or Process and Instrumentation Diagrams (P&ID), which model the system of interest and represent components of a system as nodes. A node signifies a section of the process in which a significant process change occurs. An example is illustrated in Figure 3.2. The HAZOP focuses on one node at a time. Thus, its top-down approach breaks down the analysis of risk event into manageable parts.

Upon selecting a node, the analysis begins with defining the design intent or purpose of the node and identifying parameters that evaluate the fulfillment of this design intent. HAZOP then methodically questions

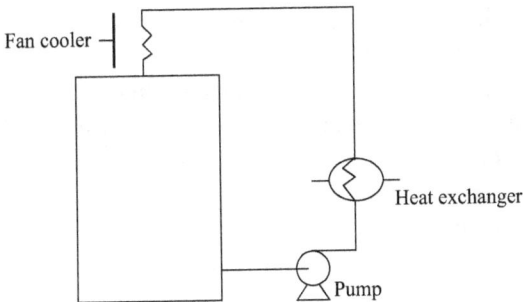

Figure 3.2. PFD of water cooling circuit pipework.[4]

every part (including materials, inputs, components, and process steps) related to the node to establish how deviations from its design intent can occur. This distinctive step of analyzing the deviation of elements of the system helps in identifying the hazards, a step that is somewhat only implicit in PHAs. This analysis is documented in a HAZOP worksheet as shown in Figure 3.3. The following are the descriptions of columns that correspond to the HAZOP worksheet. The columns also follow the general key questions of the risk management process.

- *What can go wrong?*
  - ○ Column 1: Elements associated with the node
  - ○ Column 2: Deviations from the design intent as it relates to each element
- *What are the causes and consequences?*
  - ○ Column 3: Potential causes (i.e., hazards) for the deviation to occur
  - ○ Column 4: Consequences of each potential cause or hazard
- *What are the likelihood or chances of occurrence?*
  - ○ Omitted in HAZOP
- *What can be done to detect, control, and manage them?*
  - ○ Column 5: Existing measures that prevent, control, and mitigate the hazard
  - ○ Column 6: Recommendations for additional preventive and control measures to address the hazards

3.2.3 GENERALIZED STEPS FOR HAZOP

1. Define the system and scope of study.
2. Divide the system into nodes and define each node's design intent.
3. For each node, identify node elements such as material, input, process step, or operation.
4. Define the element's relation to the design intent, performance measure, and acceptable range of performance (parameters).
5. Define the deviation by selecting a process guideword and pair with parameter.
6. Establish the causality of the deviation with potential hazards.
7. Describe the consequences of each hazard.
8. Identify existing preventive and control measures to address the hazard.
9. Determine recommendations for additional controls to address the hazard.

Node: water cooling circuit pipework

			Description			Analyst date
(1) Element	(2) Deviation	(3) Possible causes	(4) Consequences	(5) Safeguards (preventive)	(5) Safeguards (reactive)	(6) Actions required
Describe what the guideword pertains to (material, input, process step, etc.)	*What can go wrong (describe the deviation)*	*How can the deviation occur*	*What may happen if the deviation occurs*	*List controls that prevent the deviation from occurring*	*List controls that can address the deviation that is occurring*	*Identify any additional controls or actions required*
Flow process	Low flow rate No flow	Pump failure	Water backs up Production facility or equipment overheats and breaks down	Regular maintenance	Engage backup pump Replace or repair pump	Upgrade system with digital monitoring and alarm
		Broken circuit pipe	Water leak Production facility or equipment overheats and breaks down	Regular maintenance	Replace or repair pipe	Upgrade system with digital monitoring and alarm
Cooling process	High temperature	Temperature control failure	Water not cooled Production facility or equipment overheats and breaks down	Regular maintenance	Engage manual control Replace or repair controller	Upgrade system with digital monitoring and alarm
		Compressor failure	Water not cooled Production facility or equipment overheats and breaks down	Regular maintenance	Replace or repair compressor	Upgrade system with digital monitoring and alarm

Figure 3.3. HAZOP worksheet for water cooling circuit pipework.[5]

Step 1 defines the scope of the analysis and system of study. As an example for HAZOP, consider a cooling water facility for a production plant as a system of interest.[6]

Step 2 divides the system into nodes. Part of the cooling water facility is the water cooling circuit pipework as shown in Figure 3.2. This represents a node in an overall HAZOP study of a cooling water facility.

To begin the analysis of each node, the design intent of the node and its corresponding parameters are defined. For the water cooling pipework, its purpose is to continuously circulate cooling water.

For the succeeding steps, refer to the HAZOP worksheet in Figure 3.3.

Step 3 identifies the node elements. Elements are described in terms of its materials, inputs, components, and process steps and entered in Column 1 of the HAZOP worksheet. For the water cooling pipework, the key processes are circulation and cooling.

Step 4 defines process parameters for each element as it relates to the design intent. As defined in Step 2, the purpose of water cooling pipework is to continuously circulate cooling water and a key process identified in Step 3 is circulation. How does circulation then achieve the system's design intent? To continuously circulate water, the system is designed to flow at a rate of 75.5 m³/min.

Step 5 defines the deviation. It follows then that for the water cooling pipework, a deviation or departure from the design intent would be a lack of adequate circulation. In HAZOP, the term *less flow rate* or *no flow* describes this deviation. This information is entered in Column 2 of the HAZOP worksheet.

Note that HAZOP requires the use of guidewords (such as *no, more, less, high, low, reverse,* or *other than*). The guidewords are paired with parameters (e.g., volume, temperature, flow, velocity) associated with the system elements. The word combination is then used to represent a deviation from how the node (or process) is expected to function by design. Examples of paired words that represent deviations are *low volume, high volume, low temperature, high temperature,* or *no flow.* HAZOP requires exhaustive pairing of guidewords with identified parameters but not all combinations will be meaningful like *reverse volume* or *reverse temperature.*

Step 6 identifies the causes (i.e., the hazards) that can cause the deviation to occur. For the deviation *no flow,* failure of the pump may be one of the potential causes. This information is entered in Column 3 of the HAZOP worksheet. Note that the failure of the pump is described as a cause, not a deviation. Also note that in HAZOP, the identified causes are analogous to what PHA identifies as hazards.

Step 7 defines the consequences for each hazard. Pump failure means water is not flowing or circulating and, therefore, not cooled to the requirements of the production facility. A consequence could be possible overheating of equipment and interruption or delay in the production process if the work has to stop because of down equipment. This information is entered in Column 4 of the HAZOP worksheet.

Step 8 describes existing measures on how the deviations and the hazards are addressed. For the water cooling pipework, suppose a backup pump is in place, then its information is entered in Column 5 of the HAZOP worksheet.

Step 9 determines recommendations. An assessment has to be made as to whether existing measures are sufficient and effective. Part of the analysis is to identify additional safeguards or controls when and if needed. This information is entered in Column 6 of the HAZOP worksheet.

3.3 JOB SAFETY ANALYSIS

A JSA is a qualitative technique used to identify and control workplace hazards to prevent accidents. In a JSA, each basic step of the job is examined to identify and address potential hazards. The aim is to recommend the safest way to do a job. It focuses on the relationship between the worker, the task, the tools, and the work environment.[7] Similar to HAZOP, JSA does not attempt to quantify risk in terms of likelihood or consequence. Variants of JSA are job hazard analysis (JHA) and task hazard analysis (THA).

3.3.1 WHERE AND WHEN TO USE JSA

Jobs that have taken on changes in procedures, equipment, and materials are good candidates for JSA, as are jobs affected by new or revised regulations or standards. Newly created jobs, uncommon or seldom-performed jobs with little or no history, and routine jobs requiring regular or continuous exposure to hazards should undertake a JSA.

Ideally, all jobs should undergo safety analysis. If possible, JSAs should also be performed for nonroutine activities such as maintenance, repair, or cleaning and jobs that are offsite, either at home or on other job sites (e.g., teleworkers). Examining hazards to visitors and the public should also be included, as well as any special population that may need accommodations such as expectant mothers, inexperienced workers, and persons with disability.

As it involves extensive documentation and recording of job tasks, JSA is an effective setting to create and improve standard operating procedures (SOP). It is useful to train, educate, and update employees regarding job operation and safety practices. It can serve as a guide for employee performance review, training, and accident investigations.[8] JSAs are widely used in the construction, manufacturing, oil and gas, and many labor-intensive industries.

JSA can also complement PHA, HAZOP, or other risk studies by providing supporting analysis on specific job- or task-related elements, particularly those that are safety critical or ranked high priority.

3.3.2 JSA FORMAT

The tabular form or spreadsheet is the typical format of a JSA worksheet. An example[9] is shown in Figure 3.4. Unlike PHA and HAZOP, the columns of the JSA worksheet generally follows only two of the key risk questions of risk management discussed in Chapter 2: *What can go wrong? What can be done to detect, control, and manage them?* The other questions are omitted.

The JSA worksheet is intended to be completed one column at a time. The following are the descriptions of columns that correspond to the JSA worksheet shown in Figure 3.4.

- *What can go wrong?*
  - ○ Column 1: Job steps or tasks
  - ○ Column 2: Hazards or unwanted events related to each task
- *What are the causes and consequences? What are the likelihood or chances of occurrence?*
  - ○ Omitted in JSA
- *What can be done to detect, control, and manage them?*
  - ○ Column 3: Prevention and control measures for each hazard
  - ○ Column 4: Person responsible for the prevention or control measure

3.3.3 GENERALIZED STEPS FOR JSA

1. Break the job down into a sequence of basic steps or tasks.

2. Identify potential accident or hazards.
3. Determine preventive and control measures to address the hazards.

Step 1 breaks down the job into a sequence of basic tasks or steps. The listing of the steps is done sequentially according to the order in which the tasks are performed. This can be done by direct observation or the group discussion method. Observing an experienced employee coupled with enlisting other employees or a supervisor to validate the task steps can be most effective.

The use of action words or phrases (e.g., pick up, turn on) to describe each step helps in breaking down a job into steps. If observed, steps that are not part of the standardized job (SOP) should be included (and addressed for appropriateness). To support high-quality documentation, video recordings of the job in action can be compiled.

Note that the JSA worksheet should be completed one column at a time. Thus, all steps in a job must be identified and listed first, prior to proceeding to Step 2. As an example of a JSA, consider the job of welding in confined spaces. The basic tasks for the job are (1) pre-entry, (2) entry, (3) welding, and (4) completion and exit. This information is entered in Column 1 of the JSA worksheet.

For the succeeding steps, refer to the JSA worksheet in Figure 3.4.

Step 2 identifies all possible hazards related to each task. The emphasis is on identifying those that can lead to accidents and injuries for the worker, as well as potential worker errors. Hazards might exist due to the characteristics of the procedure or tasks, materials, equipment or tools used, and the work site or environment. Consider also foreseeable unusual conditions and impacts of hazards during such conditions, such as a power outage.

For the task (3) *welding*, the hazards identified are getting flashed, getting burned, getting hot metal in the face or eye, and inhaling fumes. This information is entered in Column 2 of the JSA worksheet.

Step 3 determines the recommended proposals, actions, or procedures for performing each step that will eliminate or reduce the hazard. Preventive and control measures can range from developing safe job procedures to proposing safety equipment or altering physical or environmental conditions in the job.

For the specific welding hazard—*getting burned*, a safeguard is using protective clothing such as welding jacket, mask, goggles, and gloves. This information is entered in Column 3 of the JSA worksheet. Column 4 identifies the person responsible for each measure.

Department:	Job analyzed: Welding in confined spaces	Performed by: Supervisor:	Date:
Location:		Analysis by: Reviewed by: Approved by:	Started: Completed:

Required personal protective and emergency equipment:
Gas, oxygen detector, extraction fan, blower, creeper, lifeline, hardhat, welding jacket, mask, goggles, and gloves

(1) Sequence of basic job steps *List the tasks required to perform the activity in the sequence they are carried out*	(2) Potential accident or hazards *Per task, list the hazards that could cause injury when the task is performed*	(3) Control or prevention measures *Per hazard, list the control measures required to eliminate or reduce the risk of injury arising from the hazard*	(4) Person responsible *Per control measure, identify the person responsible to implement the control or prevention measures*
Pre-entry	Pressurized fluid	Ensure the confined space is depressurized and content is drained or cleaned	Maria
	Flammable or toxic atmosphere	Ensure the confined space is flushed, purged, and thoroughly ventilated as required	
	Lack of oxygen	Ensure confined space is ventilated with sufficient oxygen for normal breathing	
Entry	Lack of contact with worker	Place attendant near entry	John
	Dark and insufficient illumination	Check lighting, replace bulbs if needed	

Step	Hazard	Controls	Responsible
Welding	Flammable or toxic atmosphere Fire and explosion	Use extraction fan or blower Monitor oxygen and gas levels Use creeper or lifeline	
	Lack of oxygen Asphyxiation	Use extraction fan or blower Monitor oxygen and gas levels Use creeper or lifeline	
	Getting flashed	Wear welding mask Change angle of weld Use nonreflective board	David
	Getting burned	Wear protective clothing, welding jacket, gloves, and welding mask	
	Hot metal or sparks getting in face or eye	Wear goggles or safety glasses under welding mask	John, David
	Fire hazard from sparks	Check for flammable materials Wear welding jacket Fire watch Check if fire extinguisher is available	
	Inhalation of fumes	Use extraction fan or blower Perform gas test Use creeper or life line	John
Exit and completion	Foreign materials left inside	Ensure there is no tool, equipment, rags, or other material left inside	John, David

Figure 3.4. JSA worksheet for welding in confined spaces.[10]

3.4 FAILURE MODE AND EFFECTS ANALYSIS

FMEA is a top-down quantitative technique applied to identify failures in a design, a process, a system, or an existing product, equipment, or service. FMEA is used to analyze elements or components of a system, their interactions and their effects on the operation of the system as a whole. The term *failure modes* pertains to ways in which functional elements may fail. The term *effects analysis* describes the consequences of those failures and any existing prevention or control measures to address them.[11] Thus, FMEA similar to HAZOP also examines how existing system capabilities detect failures and manage such failures. Correspondingly, risks are ranked based on the likelihood and consequences of failure, as well as the current system's ability to detect and manage failure if they do occur. Similar to PHA and unlike HAZOP and JSA, FMEA further assesses the likelihoods of occurrence of these failures.

3.4.1 WHERE AND WHEN TO USE FMEA

FMEA is used to study a variety of systems and products that are more detailed and complex such as those in the aviation and electronics industry. It is applicable in both the design and operation phases, particularly in the development of operational risk management strategies. It is useful for redesigning existing systems for a new application as well. FMEA is also used for risk and safety auditing and accident investigations. More often, it is applied as periodic review throughout the life of the system to accommodate changes within the system and its environment.

FMEA is frequently used in combination with other tools such as FTA or CCA. An example of this combined application is discussed in a later section in this chapter.

3.4.2 FMEA FORMAT

A tabular form or worksheet is used to collect information and document failure analysis in FMEA. Figure 3.6 shows an example[12] of an FMEA worksheet. Similar to HAZOP studies, the FMEA worksheet is often done in reference to a larger system analysis. Correspondingly, each FMEA worksheet typically refers to one of the parts or components depicted in a system diagram such as PFD or P&ID used in the application of HAZOP. An item analysis for simple systems may also be used, as illustrated in Figure 3.5. An item analysis is a listing of a system's basic components.

The FMEA worksheet generally follows the key questions of the risk management process. Very notably, the precursory question *What should go right?* is posed first. It is similar to the notion of *design intent* in HAZOP. However, in FMEA, the design intent is explicitly documented and captured in the worksheet itself, unlike in HAZOP where it is primarily implied. The following are the descriptions of columns that correspond to the FMEA worksheet shown in Figure 3.6.

- *What should go right?*
  - Column 1: Functional objectives of the subsystem or component
- *What can go wrong?*
  - Column 2: Failure modes that prevent the achievement of objectives
- *What are the causes and consequences?*

The failure modes are described as risk events in terms of
  - Column 3: Consequences of failure (labeled *effects* in the FMEA worksheet)
  - Column 4: Severity (S) of the consequences
  - Column 5: Potential causes of failure
- *What are the likelihood or chances of occurrence?*
  - Column 6: Likelihood (L) of each causes or failure (labeled *occurrences* in the FMEA worksheet)
- *What can be done to detect, control, and manage them?*
  - Column 7: Existing detection, prevention, and control measures
  - Column 8: Rating for existing detection (D), prevention, and control measures
  - Column 9: The risk score determined by the likelihood, severity, and effectiveness of existing detection, prevention, and control measures to address failure, evaluated as $L \times S \times D$
  - Column 10: Recommendations for additional prevention and control measures to address the failures, their causes, and consequences

3.4.3 GENERALIZED STEPS FOR FMEA

1. Define the system and the scope of the study.
2. Identify the key functional objectives.
3. For each objective, identify the ways failure can occur (the failure modes).

4. For each failure mode, identify the consequences (effects) if failure occurs.
5. Rate the consequences in terms of severity.
6. Identify the root causes for each failure mode.
7. Rate each root cause in terms of chance of occurrence.
8. Identify current preventive and control measures for each root cause.
9. For each root cause and failure, rate how well or early they can be detected and addressed.
10. Calculate the risk priority number (RPN) based on the consequences, occurrence, and detection.
11. Identify the failures that are considered critical.
12. Determine additional preventive or control measures as needed.

Step 1 identifies the system and the scope of the study. FMEA can be a comprehensive look at the total system as a whole, or it may focus on certain subsystems, elements, or components. As an example, consider the *bicycle* as a system of interest. An item analysis of the bicycle system is shown in Figure 3.5.

The FMEA worksheet in Figure 3.6 focuses on a particular subsystem, the *hand brake subsystem*, and refers to Item 1.8 (shaded in gray) in the item analysis of the bicycle in Figure 3.5.

ITEM ANALYSIS—BICYCLE EXAMPLE

1.0 Bicycle System   ←*System item*

   1.1 Frame subsystem
   1.2 Front wheel subsystem
   1.3 Rear wheel subsystem
   1.4 Sprocket–pedal subsystem
   1.5 Chain–derailleur subsystem
   1.6 Seat subsystem
   1.7 Handle bar subsystem

   1.8 Hand brake subsystem   ←*Subsystem item*

     1.8.1 Brake lever
     1.8.2 Brake cable   ←*Component item*
     1.8.3 Brake pads
     1.8.4 Brake caliper

   1.9 Suspension subsystem

Figure 3.5. Item analysis of bicycle.[13]

FMEA

Item No.: 1.8

Item description: Hand brake subsystem

Date:

(1) Function or process	(2) Potential failure modes	(3) Potential effect(s) of failure	(4) Severity (S)	(5) Potential cause(s) of failure	(6) Occurrence rating (L)	(7) Current design controls (prevent)	(7) Current design controls (detect)	(8) Detection rating (D)	(9) RPN	(10) Recommended action(s)
What should go right, what is supposed to happen	*What are the ways that prevent the process or function to achieve its objective*	*What is the consequence of the failure*	*How significant is the impact*	*How can the failure occur*	*What is the likelihood of occurrence*	*What are in place so that failure is prevented*	*What are in place so that failure is controlled*	*How effective are the measures*	$L \times S \times D$	*What can be done*
Provide the correct level of friction between brake pad assembly and wheel rim	Insufficient friction between brake pads and wheels during normal operating conditions or extreme conditions	Bicycle wheel does not slow down when the brake lever is pulled potentially resulting in injury	5	Cable binds due to inadequate lubrication or poor routing	2	Make sure hand brake is consistent with design guide	Make sure testing satisfies established bicycle system durability tests	1	10	Review manual
				Foreign material reduces friction between brake pads and wheel rim	1					
				Cable breaks	3					

Figure 3.6. FMEA worksheet for bicycle brake component.[14]

For the succeeding steps, refer to the FMEA worksheet in Figure 3.6. Step 2 identifies the key functional objectives of the system, subsystem, or component of interest. It describes what the component is supposed to do to achieve the functional objective—*description of what needs to go right*. Thus for the bicycle, one of its critical functions is the ability to stop safely within a required distance under all operating conditions. This function is achieved through the hand brake by providing the correct level of friction between the brake pad assembly and the wheel rim when the brake lever is pulled. This information is entered in Column 1 of the FMEA worksheet. It describes what the hand brake subsystem is supposed to do to achieve the function of stopping or braking (*What should go right?*).

Step 3 identifies the failure modes. As Column 1 defines what the component is supposed to do, Column 2 describes what can happen (*What can fail?*) that will prevent the component from doing what it is supposed to do. In other words, the failure modes are the different ways the component can fail to accomplish its functional objective. For the bicycle, one potential failure mode is having insufficient friction between the brake pad and wheel rim during normal or extreme conditions such as heavy rains. This information is entered in Column 2 of the FMEA worksheet. Note that it is important to describe in detail the conditions in which the failure mode can occur.

Step 4 identifies the consequences (effects) on the system as a whole, its elements, and the environment for each failure mode. It helps to ask *What happens when this failure occurs? What does the user experience because of this failure?* For the bicycle, insufficient friction means the bicycle wheel does not slow down when the brake lever is pulled, potentially resulting in collision, loss of control, and eventual injury. This information is entered in Column 3 of the FMEA worksheet.

Step 5 provides a severity rating for each effect. Rating scales such as those used in Chapter 2 can be used. For the bicycle, the worse potential result of the identified failure mode is safety failure; the severity is evaluated as catastrophic (see Table 2.2). A level 5 is entered in Column 4.

Step 6 identifies all potential root causes for each failure mode. What specifically causes this failure mode to occur? Here it is important to establish the causal chain of events to identify root causes. For the bicycle example, there are three potential causes for having insufficient friction between the brake pads and wheel rim—(1) failure of cable binds due to inadequate lubrication or poor routing, (2) foreign material on brake pad or wheel rim that reduces friction, (3) cable itself breaks, and (4) brake lever misalignment. This information is entered in Column 5 in the FMEA worksheet.

Step 7 rates the root causes according to their likelihood of occurrence. This rating evaluates the chance of failure because of a particular root cause. It follows that if a particular root cause is more likely to occur, then failure from this root cause is also more likely. For the bicycle, the cable breaking is the most likely cause of insufficient friction among the causes. Using Table 2.3, its likelihood of occurrence is moderately possible and given a rating of 3. This information is entered in Column 6 in the FMEA worksheet.

Steps 8 and 9 describe and rate the effectiveness of preventive and control measures that currently exist. Each measure is evaluated according to how well or early the failure mode or its cause can be detected and addressed before the effects occur. Detection is rated on a scale from 1 to 5, where 1 means the control is absolutely certain to detect the cause or failure and 5 means the inability for detection.

For the bicycle example, detection of the cable binding due to inadequate lubrication or poor routing is given a rating of 2. This is due to the quality and testing process in the manufacturing of bicycles. Table 3.2 can be used to determine a detection rating. The information for steps 8 and 9 are entered in Column 7 and Column 8 of the FMEA worksheet, respectively.

One can also evaluate reactive measures. Specifically, how the system is able to detect failure after the cause has already happened; and how the system is able to address the failure before the impact of the effects occurs. For example, if the bicycle had an indicator or an alarm that the hand brake subsystem has failed. This measure can inform the user to use alternative means for stopping or breaking.

Table 3.2. Example of detection scale[15]

Likelihood of detection	Criteria: likelihood of detection by design control	Rank
High to absolute uncertainty	No current design control or cannot be detected	5
Remote	Detection controls have a weak detection capability	4
Low	Product verification with pass or fail testing	3
Medium	Product verification with test to failure testing	2
High	Product verification with degradation testing	1

In step 10, upon identifying the likelihood (L), consequences (S) of failure modes, and the current system's ability to address (D) the causes of failure, failure modes are then evaluated for criticality using a risk score similar to PHA. For FMEA, the risk score is known as the RPN[16] and can be derived by multiplying $L \times S \times D$, corresponding to Columns 4, 6, and 8 respectively. The RPN score entered in Column 9 can provide guidance for ranking potential failures in the order they should be addressed.

Step 11 describes recommendations and proposed additional preventive or control measures if necessary. Particular attention is given to proposals that will decrease the likelihood of occurrence, lower the severity of the consequences, or increase the capability or effectiveness to detect early and mitigate the causes of failure. Part of the recommendation should include the organizational functions responsible for these actions and target completion dates. This information is entered in Column 10 of the FMEA worksheet.

3.5 FAULT TREE ANALYSIS

FTA is a deductive top-down modeling technique used to analyze how a failure can occur. It maps out causal chain of events by connecting hazards and events that can bring about a total or partial system failure. Unlike other risk tools that examine single events or single hazards at a time, FTA is able to analyze relationships of more than one events together and determine combined effects. Thus, FTA is used extensively in reliability and safety engineering to understand how systems can fail.

3.5.1 WHERE AND WHEN TO USE FTA

FTA is useful in understanding and managing larger and more complex systems. It can be used to show compliance with system safety and reliability requirements or support more robust system design. It is useful for developing system or product specifications and for showing critical contributors and pathways to system failure. FTA can extend the analysis of other risk tools such as FMEA, HAZOP, and CCA.

3.5.2 FTA FORMAT

FTA uses graphical methods or diagrams to represent failure events that can occur and show connections that define pathways to system failure.

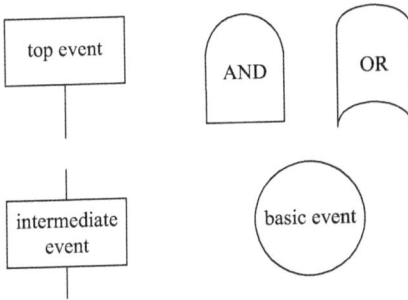

Figure 3.7. Standard symbols used in FTA.

Failure events in FTA like FMEA are described as risk events. Risk events can be conceived by asking the question *What can go wrong?*

The *top event* represents the state of the system taken to be the top risk event. Often it refers to a state of total system failure. Intermediate events represent events that may lead up to the top event. It can refer to subsystem failures or risk events that can cause the top event to occur. Events are denoted by rectangle shapes. Figure 3.7 shows the standard symbols used in FTA.

The events interconnect with standard logic gate symbols that represent dependency or contingency relations between events. Thus, two events connected underneath with the logic AND gate are events that are both necessary conditions (*both* have to happen) to cause the above event to occur. In contrast, the logic OR gate connecting two events represents two conditions that can cause the above event to occur. In this case, it takes *only one* of the two events to occur for the top event to happen.

Extending the elevator door example used in the PHA example, sensor failure and mechanism failure are two failure events that contribute to the elevator door closing on an obstruction. The logic diagram is shown in Figure 3.8. As the risk event is caused by any one of these events happening (only one has to occur), the two failure events, both located underneath the logic gate are connected by an OR gate. If either the sensor fails or the mechanism fails or both sensor and mechanism fails, the door will close on the obstruction.

Following the decomposition of the top event to intermediate events, intermediate events are further broken down to sublevels until one reaches the basic events that represent the elementary components (or root causes) that can cause the system to fail. Basic events are denoted by circles.

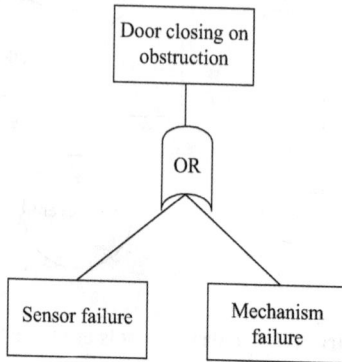

Figure 3.8. Logic diagram for the elevator example.

With the decomposition of events down to the basic events, a tree diagram is formed that outlines the different failure pathways that could cause the top event to occur. Figure 3.9 is an example of a fault tree diagram in FTA. In the traditional solution of fault trees, minimal cut sets can be determined. Minimal cut sets are all the unique combinations of events that lead up to the top event. Events that are common to these distinct paths should be identified and considered critical. The FTA model can further help in determining which pathways are most likely to occur if data is available on the likelihood of occurrence of the events.

Fault tree construction can be exhaustive, extensive, and time consuming. It is important that one focuses on those events that are regarded as critical and controllable. At times, it may make sense to identify other events that are uncontrollable or not critical and still add them to the model for completeness even if they are not further decomposed.

FTA follows a standardized method with rules (i.e., Boolean logic rules). One rule is that there should be no gate-to-gate connections. For example, an OR gate cannot be directly connected to an AND gate. Another rule is that the tree is formed level by level, which means one must exhaustively identify same level events first before constructing the level underneath.

3.5.3 GENERALIZED STEPS FOR FTA

1. Define the system to be analyzed.
2. Describe the top-level risk event (e.g., system failure).

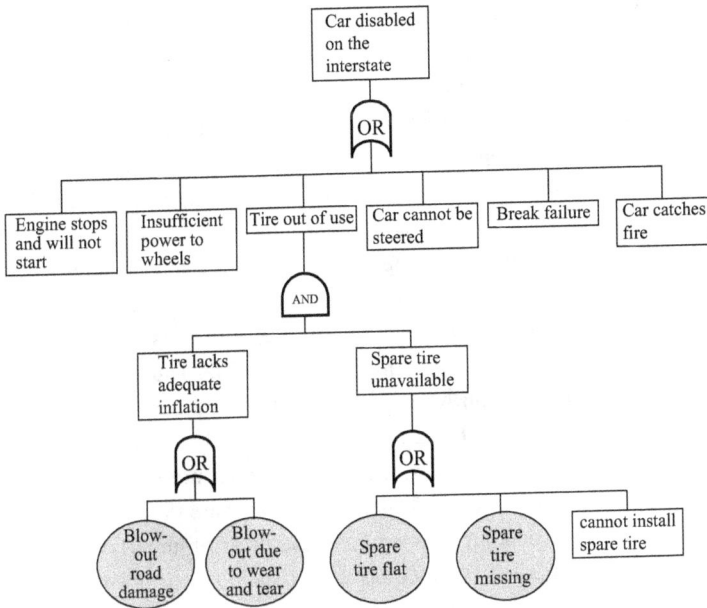

Figure 3.9. FTA for a car disabled on the interstate.[17]

3. Describe intermediate events (e.g., subsystems failure) that contribute to the top-level event and connect them with logic gates as appropriate.
4. Identify sublevel events that contribute to intermediate events and connect same level events with logic gates as appropriate.
5. Repeat (4) until the underlying contributory events are regarded as root causes (basic events).
6. Identify critical events and pathways to failure.

Step 1 begins with defining the system to be analyzed. Consider an FTA of a disabled car.[18] For the succeeding steps, refer to the fault tree diagram in Figure 3.9.

Step 2 describes the top-level risk event. In Figure 3.9, the top-level event of concern is *car disabled on the interstate*. Specific details such as the location, for example, *on the interstate*, is significant because a car disabled while in a driveway will have differing analysis. Hence, the conditions surrounding the failure should be described in as much detail as possible.

Step 3 describes the intermediate events that contribute to the top event. Below the top level are the different subsystems failures or risk events that can occur and which are potential causes to why a car is disabled in the interstate. These risk events are denoted as the intermediate events—(1) engine stops and will not start, (2) loss of power, (3) tires out of use, (4) no steering, (5) break failure, or (6) car on fire. If *at least one* of the intermediate events occurs, it can potentially cause the top-level event to happen. Thus the events are interconnected by an OR gate.

Step 4 describes the next sublevel events that contribute to the intermediate events. After identifying the intermediate events, the events can further be broken down in the next sublevel. Consider the intermediate event *tires out of use*. Two events have to happen together to cause the car to be disabled on the interstate due to tires out of use—(a) tires lack adequate inflation and (b) spare tire is not available. Thus the two events are interconnected with an AND gate. Note that in the event that the tire lacked adequate inflation but a spare tire is actually available, only one contributory event occurs. Thus, the top event does not occur and the car is not disabled because of the availability of the spare tire.

Step 5 further decomposes sublevel events until underlying contributory event are regarded as *root causes*, also called *basic events*. The root causes for the tires lacking adequate inflation are tire blow out from either *road or external damage* or *wear and tear*. The root causes of spare tire being unavailable are *spare tire flat* or *spare tire missing*.

Step 6 identifies the critical events and pathways to failure.

Thus, combinations of the basic events identified in step 5 will potentially cause the top event (car disabled on the interstate) to occur from tires out of use. These combinations are listed as follows:

1. Tire blow out from road or external damage and spare tire is flat
2. Tire blow out from road or external damage and spare tire is missing
3. Tire blow out from wear and tear and spare tire is flat
4. Tire blow out from wear and tear and spare tire is missing

3.6 CAUSE AND CONSEQUENCES

CCA is a modeling technique that uses graphical methods and diagrams. It is an analytical method for tracing and exposing the chains of events related to a particular risk event of interest. CCA serves as a visual chronological description of failures from its initiation to the final outcomes.[19]

3.6.1 WHERE AND WHEN TO USE CCA

CCA is an effective tool for illustrating the relationships between causes and consequences, especially when examining complex causal event chains where many possible causes and consequences connect to a single event. It is commonly used for analysis of security and safety problems. It can also be useful for determining requirements for risk management in the design and development of a new system. It can be used for assessing an existing system's performance standards, risk management strategies, and accountability. CCA can be applied as a stand-alone tool in managing risk as well as a support for other tools such as FMEA and HAZOP.[20]

3.6.2 CCA FORMAT

CCA, similar to FTA, uses diagrams to represent failure events and their connections that define pathways to a particular risk event of interest. However, unlike FTA, CCA is organized according to a timeline. The simplest version of a CCA is structured like a bowtie shown in Figure 3.10 where the risk event (the primary event) is positioned in the middle of the bowtie.

The sequence of events that are considered causes is drawn to the left of the bowtie. These events are initiating events that lead up to the primary event. The initiating events can be drawn as an FTA where several levels and sublevels of causality are depicted and where the top event is the primary event. As described in the earlier section, FTA is useful to show how various events may combine to lead up to the primary event.

The sequence of events that are considered consequences is drawn to the right of the bowtie. These events occur as a result of the primary event and are known as the damage chain. Damage chains can be drawn as an event tree analysis (ETA) to enumerate the possible follow-up risk events (consequences) that may arise from the primary event. An example of an ETA is illustrated in Figure 3.11.

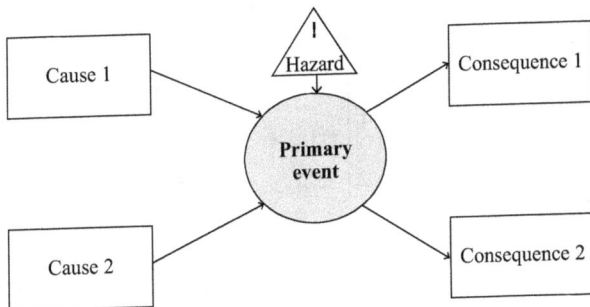

Figure 3.10. Bowtie diagram for CCA.

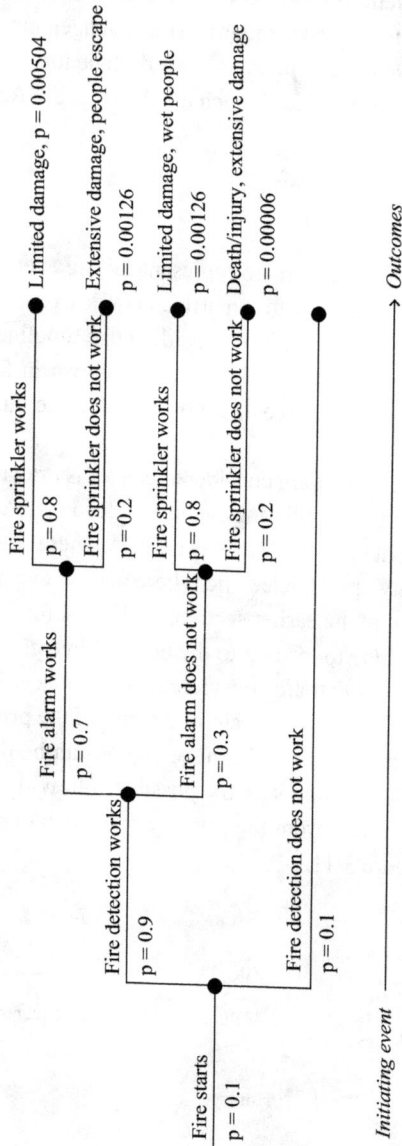

Figure 3.11. Event tree analysis.[21]

ETA is a forward logical modeling technique that explores possible consequences or outcomes through a single primary event. The event tree diagram models all possible pathways from the primary event and depicted as branches. Each follow-up event can be intermediate events that in turn initiate other follow-up events.

For example, consider the primary event *a fire starts*. The follow-up events that are considered are (1) fire detection, (2) fire alarm, and (3) fire sprinkler. For each follow-up event, possible outcomes are enumerated. For fire detection, two things can possibly happen—*the fire detection works* or *fire detection does not work*. Note that in this example, each succeeding event has two possible mutually exclusive outcomes that can occur. In many cases, there may be more than two outcomes.

CCA can be used fairly directly for quantification of risk and probabilities, but the diagrams can become exhaustive and cumbersome. Because of this, CCA is not as widely used as FTA or ETA. Fault and event trees are easier to follow and so tend to be preferred for presentation of the separate parts of the analysis.[22]

3.6.3 GENERALIZED STEPS FOR CCA

1. Define the system of interest.
2. Identify the primary event.
3. Generate the initiating chain (similar to FTA if applicable).
   - Determine the underlying causes of the event.
   - For each underlying cause, identify the causes or initiating events.
   - Repeat until the underlying cause becomes uncontrollable.
4. Generate the damage chain (similar to ETA if applicable).
   - Determine follow-up events.
   - For each follow-up event, identify the consequences.
   - Repeat until all outcomes are exhausted.

Step 1 identifies the primary event. Consider an FMEA worksheet of a brewing process shown in Figure 3.12. The FMEA defines the primary event from one of its failure modes, bad batch (shaded in gray). The CCA focuses on this primary event as illustrated in Figure 3.13. As shown in the example, the CCA can be used to extensively detail the causes of failure and their effects.

Item No.: XXX
Item description: brewing process
Date:

(1) Function name	(1) Function ingredients used	(1) Function equipment used	(1) Function process description	(2) Potential failure modes	(3) Potential effect(s) of failure	(5) Potential cause(s) of failure	(7) Control measures
Mashing—converts starches released during malting into sugars for fermentation through saccharification. This step produces the wort	Malt and mash source (malted barley and assorted grains called the grain bill)	Mash tun	1–2 hour process where water and grain bill are heated inside the mash tun	Starches of grain bill are not fully converted to sugars	Lower yield of wort batch is scrapped	Operator error, improper process, equipment malfunction, bad ingredients	SOP, training, quality control
Fermentation—sugars generated in wort are converted into alcohol and CO_2, producing unfiltered and unconditioned final product	Brewer's yeast and flavored hops	Fermentation vessels	Fermentation vessels are used to store mixed wort and add brewer's yeast or other flavorings. Yeast flocculate and precipitate to bottom of the vessel when fermentation is finished	**Bad batch**	Bad batch is scrapped; Bad batch is distributed	Operator error, improper storage, bad ingredients	Recipe sheet, training, quality control, temperature, and humidity-controlled storage

*Note that for this FMEA example, Columns 4 and 6 are omitted for simplicity.

Figure 3.12. FMEA of brewing process.

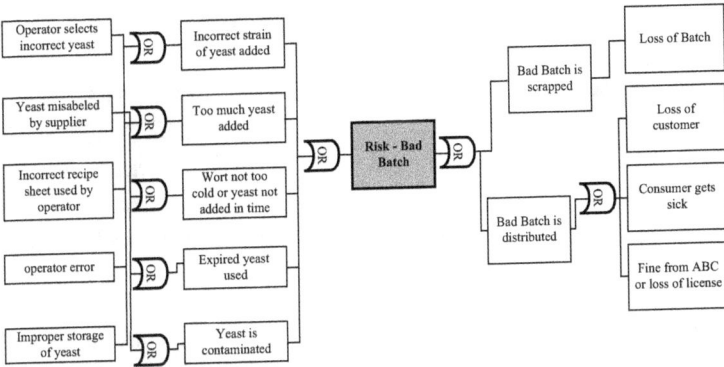

Figure 3.13. CCA of bad batch.

Step 2 generates the initiating events. The bowtie structure is applied as a simplified form of CCA. Underlying causes of a bad batch include incorrect strain of yeast added, too much yeast added, wort too cold, and expired or contaminated yeast.

Consider the first cause—*incorrect strain of yeast added. How can this happen?* It can happen when (1) the operator selects the incorrect yeast or (2) the supplier mislabels the yeast. The initiating events are ordered sequentially from left to right and events that cause other events are connected by links such as those used in an FTA.

Step 3 generates the damage chain. The consequence of a bad batch is either it is scrapped or it is distributed, which can result in loss of customer, consumer getting sick, fines, or loss of license.

3.7 AS LOW AS REASONABLY PRACTICABLE

ALARP is a principle and not a technique like those described in the earlier sections of this chapter. The ALARP principle supports the notion that residual risk shall be as low as reasonably practicable. Since safety (i.e., zero risk) cannot be guaranteed and the cost of addressing risk can be excessive if pursued without limits, ALARP provides an approach to help derive a justifiable level of riskiness and set tolerable risk levels. The approach involves judgment in weighing the riskiness and the cost of attaining such levels. Thus, the concept also includes accepting a certain level of risk because of the deliberate decision to bear it rather than address it. The ALARP principle is also similar to SFAIRP (so far as is reasonably practical).[23]

3.7.1 WHERE AND WHEN TO USE ALARP

ALARP is used primarily to address the questions *What can be done to detect, control, and manage them? What are the alternatives?* and *What are the effects beyond?* Given that a risk study has been undertaken, whether through an FMEA, HAZOP, PHA, or other techniques, decision makers are then presented with recommended or proposed measures to treat the particular risk events. At this point, decision makers need to decide if a certain risk is to be addressed. If it is, how much of the risk should be addressed. Or if it is not, whether the risk can be reasonably accepted or tolerated. ALARP is a way of thinking that aids making such decisions.

The approach can also be applied preemptively during the recommendation step of risk studies. When analysts are cognizant of ALARP, they should propose measures that already adhere to a level ALARP.

3.7.2 ALARP FORMAT

In ALARP, "reasonable practicability" involves weighing a risk against the benefits. And "low" suggests a reduction of risk to a level wherein the benefits arising from further risk reduction are disproportionate to the time, cost, and technical difficulty of implementing further risk reduction measures.[24] For example, is it worth to spend $30 on a bicycle helmet to gain an 88 percent reduction in the risk of severe brain injury from riding a bike? If one chooses to make the investment, then one chooses to reduce the risk to this ALARP level. In this case, the benefit of a reduction of risk from a bike injury is worth (or even outweighs) the cost and effort of $30 and a trip to the store. Thus, the risk of injury from biking *without* a helmet is *not* ALARP. While, the reduction of risk from biking *with* a helmet *achieves* a risk that is ALARP. Note that even with biking *with* a helmet and its corresponding risk reduction, the risk from biking is still not totally eliminated.

Conceivably, one can choose not to go biking at all, which would theoretically even further reduce the risk of head or brain injury *from biking* to zero. But does one forgo the exercise and entertainment provided by biking against this risk reduction of not engaging in the activity? Those who like biking will conclude that this particular risk reduction is *not* ALARP. The benefit is *not worth* the cost (of not experiencing the joys of bicycling). However, one can argue that there will be extremes, the few who choose not to go biking because of the risk. For them, the choice not to go is ALARP. As can be perceived, ALARP tends to be relatively defined, inherently subjective, and value laden.

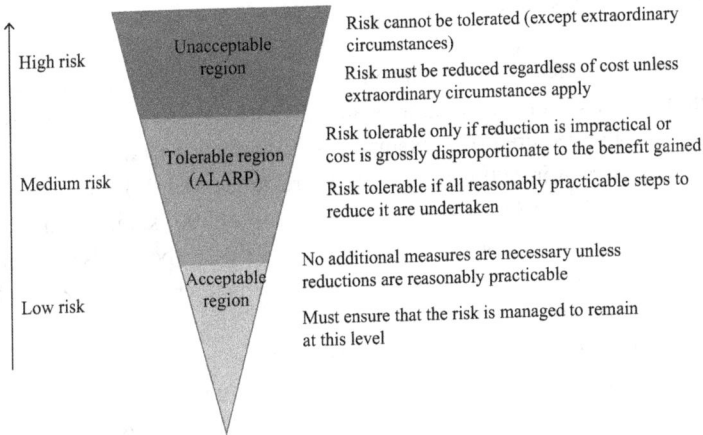

		Risk cannot be tolerated (except extraordinary circumstances)
High risk	Unacceptable region	Risk must be reduced regardless of cost unless extraordinary circumstances apply
Medium risk	Tolerable region (ALARP)	Risk tolerable only if reduction is impractical or cost is grossly disproportionate to the benefit gained
		Risk tolerable if all reasonably practicable steps to reduce it are undertaken
Low risk	Acceptable region	No additional measures are necessary unless reductions are reasonably practicable
		Must ensure that the risk is managed to remain at this level

Figure 3.14. Carrot diagram of the ALARP principle.[25]

The carrot diagram in Figure 3.14 illustrates the ALARP approach and is often used to display levels of risk. The term carrot diagram comes from its appearance as an elongated triangle, which looks like a carrot.

At the top are the high (normally unacceptable) risks. Risks that fall in the unacceptable region must be reduced regardless of cost, except in extraordinary circumstances. These risks are definitively *not* at ALARP levels. They characteristically include those risks that may lead to death or disability. An example of a risk in this region is not outfitting a car with seatbelts and airbags or window washing in skyscrapers without harnesses. As can be expected, many risks that fall under the unacceptable region are those that are mostly controlled by laws, regulations, and codes. Industry practices are good references as well for determining common preventive and control measures that achieve ALARP levels. Other significant risks that should be identified in this region are those that may threaten the core mission of the organization. Addressing these risks that are vital to the survival and sustainability of a business should be reduced at all costs.

The low risks fall at the bottom. These risks that fall in the broadly acceptable region are risks that are tolerable. They are more likely to be at ALARP levels. For example, the risk of being pickpocketed is so low that people do not feel the need to carry cash in separate pockets or hidden money belts. Similarly, we manage slightly higher risks, such as crossing a neighborhood road by routine procedures (look left and right) that we were taught as children. Looking before crossing the road reduces the risk to an individual's ALARP levels. Careful monitoring and attention must be done to ensure that risks in this region remain at this level.

3.7.2.1 Finding the (Im)Balance

If the costs are clearly very high and the reduction in risk is only marginal, then it is likely that the situation is already ALARP and further improvements are not required. For example, is it worth to spend $1 million to prevent 1 person in 500 getting a small bruise per year? If the answer is no, then the risk of injury (bruise) of 1 person in 500 is deemed already low as reasonably practicable (at a level ALARP). The benefit is not worth the cost and effort. The risk event of the bruise is tolerated. This risk falls in the low risk acceptable region.

In other circumstances, the improvements may be relatively easy to implement and the risk reduction is significant: Here, the existing situation is unlikely to be ALARP prior to the improvement, and thus the improvement is performed (such is the case with the bicycle helmet example). In many of these cases, a decision can be reached without further analysis. The difficulty is when the risk falls between the upper and lower limits, the tolerable region, when the difference between the risk and the benefit are not grossly disproportionate.

Studies using Benefit–Cost (BC) analysis have been used to support these types of ALARP decisions. A limitation of BC is that it assumes all factors involved can be converted to monetary values. Unfortunately, it is well known that there are estimating issues and implied value judgments in appraising or assigning monetary values for both benefits and costs. This is found particularly difficult where the value of human life and the (assigned) cost of suffering and deterioration of the quality of life play a major role in the analysis. Similar difficulties arise in the valuation of environmental and social impacts.

Decision-making models are also useful for supporting ALARP. Multicriteria decision making (MCDM) or multicriteria decision analysis (MCDA) considers multiple criteria other than cost. MCDM and MCDA are able to integrate factors that need not be assigned cost valuations such as human life or environmental degradation. It allows criteria to preserve its intrinsic unit of measure (e.g., mortality rate, pollution level). Nonetheless, even if cost conversion is circumvented, the exercise of valuation (e.g., quantifying life) still exists in many decision-making models.

3.7.3 GENERALIZED STEPS FOR ALARP

1. Identify risk.
2. Determine where risk falls in the ALARP region.

3. If risk falls in the top region, undertake implementation plan to achieve ALARP.

4. If risk falls in the bottom region, implement proposals if simple and inexpensive; undertake monitoring and control to maintain ALARP.

5. If risk falls in the middle region, undertake further analysis (e.g., BC analysis, MCDM, or MCDA).

3.7.4 ALARP IN PRACTICE

Many existing industry standards and "industry good practices" are already available for risk management practices. These are widely accepted as ALARP in the industry. Survey literature and information for

- Health and safety guidelines
- Industry standards specification
- International, federal, and state standards, laws, and regulations
- Suggestions from advisory bodies
- Comparison with similar hazardous events in other industries

As knowledge and technology develop, new and better methods of risk control become available. A review of what is currently available should be studied and considered. Note that the ALARP approach should still be applied. The latest and even the best risk controls available are not necessarily reasonably practicable for a particular organization.

In fact, in some cases, a risk in the unacceptable region may not be feasibly addressed because the organization or business simply does not have the resources to reduce the risk. For example, if the cost of reducing a risk puts the business in a significant financial loss and possibly bankruptcy, then the business will likely choose to bear the risk instead. It can still be argued that ALARP was achieved since that level of risk (even though high) was what was reasonably practicable for that particular organization. Again, this exposes the subjective nature of ALARP.

Thus ALARP entails significant stakeholder input to ensure support and establish accountability. A process of discussion with stakeholders can be done to achieve a consensus about what is ALARP. Public perception may need to be considered as well. For example, in policy work, one may also need to consider how the public feels about a certain risk. It may be important to gauge public perception and incorporate the information in the risk analysis.

Table 3.3. Summary comparison of tools or techniques and the principle described in this chapter

Name of the techniques	What it is?	What is the underlying purpose?	Where to apply it?	How to apply it?
PHA	Initial assessment of the hazards and their corresponding accidents	To identify the hazards, hazardous situations, and events that can cause harm for a given activity, facility, or system	An initial risk study in an early stage of a project; or an initial step of a detailed risk analysis of a system concept or an existing system	PHA worksheet
HAZOP	A structured and systematic qualitative technique for system and operability assessment	To systematically identify possible *deviations* from the design and operating intent and determine how these deviations may eventually become hazards	General-purpose risk management tool. It is applicable to various types of systems, especially those with highly monitored performance and detailed design requirements	Use HAZOP worksheet Extend with JSA, FTA, or CCA where applicable
JSA	A qualitative technique used to identify and control workplace hazards to prevent accidents	To identify hazards associated with each step of any job or task that has the potential to cause serious injury	Workplace hazards Should be performed before every job	Use JSA worksheet Video record job if possible
FMEA	A structured and systematic quantitative technique applied to identify failures in a design, a process, a system, or an existing product, equipment, or service.	To analyze elements or components of a system, their interactions, and their effects on the operation of the system as a whole	General-purpose risk management tool. It is applicable to various types of systems, especially those with highly monitored performance and detailed design requirements	Use FMEA worksheet Extend with JSA, FTA, or CCA where applicable

(Continued)

Table 3.3. Summary comparison of tools or techniques and the principle described in this chapter (*Continued*)

Name of the techniques	What it is?	What is the underlying purpose?	Where to apply it?	How to apply it?
FTA	A deductive top-down modeling and graphical technique used to analyze how a failure can occur	It maps out causal chain of events by connecting hazards and events that can bring about failure. Unlike other risk tools that examine single events or single hazards at a time, FTA is able to analyze relationships of more than one event together and determine combined effects	General-purpose risk management tool. It is useful to understand and manage larger and more complex systems to show compliance with system safety and reliability requirements or support more robust system design	Boolean logic Fault tree diagram
CCA	A graphical technique for analyzing cause and consequence chains	To show the way various factors may combine to cause a hazardous event, and to show the various possible outcomes of these events	Can be applied especially when examining complex event chains with multiple chains	Bowtie diagram FTA ETA
ALARP	A fundamental approach that sets the risk to the tolerable, reasonable, and practical level	To reduce the risks to the reasonable level at which the benefits arising from further risk reduction are disproportionate to the time, trouble, cost, and physical difficulty of implementing further risk reduction measures	When evaluating whether to implement proposals, control, and preventive measures	Carrot diagram Cost–benefit analysis if needed

Finally, we have to realize that the hazard or risk *may actually occur* at some time despite all the efforts and resources put in to address it (recall that many risks cannot be reduced to absolute zero). This is an uncomfortable thought for many people, but it is inescapable.

3.8 SUMMARY

The tools and techniques discussed in this chapter are PHA, HAZOP, JSA, FMEA, FTA, and CCA. The principle of ALARP was also described to give guidance in applying the risk management process.

Table 3.3 summarizes and compares the common tools or techniques and the principle.

CHAPTER 4

TREATMENT OF RISK

4.1 RISK TREATMENT IN MANAGING OPERATIONAL RISKS

Risk treatment (including risk mitigation) is a general term to indicate modifying a system or its elements for the purpose of affecting particular risk events. Risk treatment follows the identification and description of risks, the latter also termed as risk assessment.

4.1.1 WHY RISK TREATMENT IS PERFORMED?

Risk treatment is performed or considered for implementation when particular risks are deemed important enough that actions need to be taken. Importance is usually based on the chances of occurrence, the consequences if the risk events happen, and other factors such as ethical and legal reasons.

4.1.2 WHAT ARE THE OBJECTIVES OF RISK TREATMENT?

The general objectives in risk treatments are (a) reduction in the chances of occurrence of the risk events and (b) reduction in the consequences if the risk events occur.

4.1.3 HOW ARE CHANCES OF OCCURRENCE REDUCED?

Reducing the chance of occurrence, also known as prevention, involves preventing the causal chain of events (or causalities) to develop or progress. Part of risk analysis is establishing the causal chain of events that eventually lead to a particular risk event under consideration, such as those embodied in Failure Mode and Effects Analysis (FMEA), Hazard and Operability Analysis (HAZOP), and Cause and Consequences Analysis (CCA) as discussed in Chapter 3. Each discernable segment in the causal chain of events may serve as an opportunity to terminate the progression of the risk event, hence reducing the chance of its eventual occurrence.

4.1.4 HOW ARE UNDESIRED CONSEQUENCES REDUCED IF THE RISK EVENTS OCCUR?

Once a risk event occurs, the accrual of the consequences may be spread out over a time period rather than an instantaneous accrual. The general objective is to slow down the accrual of the consequences and to eventually stop it as soon as possible. This is also known as risk mitigation.

4.2 FUNDAMENTAL STRATEGIES OF RISK TREATMENT

In risk treatment, there are various strategies for (1) reducing chances of occurrence, (2) reducing consequences if they do occur, or (3) both. Detection and control are the typical strategies to reduce the chances of occurrence and are often applied in anticipation of a risk event, while recovery plans address the reduction of consequences after risk events have occurred. As an example, consider the risk event of a fire that may result in injury, death, or damage to property and, in an industrial setting, may include the disruption of operation. To make risk treatment more concrete, as an example, consider an industrial fire scenario in which the causalities may follow the sequence shown in Table 4.1.

4.2.1 DETECTION

Detection pertains to the capability to obtain information on the presence or absence of a particular event or phenomenon, especially of events causing, and precursory to the risk events of interest (the risk event of interest is from here on termed as the *target risk event*).[1] Early and accurate detection of the causal chain of events provides opportunities to prevent or control succeeding events to happen, essentially breaking the causal chain of events that lead to the target risk event and further progression of the sequence of events and accrual of consequences.

Table 4.1. Example of the causal chain of events leading to injury and equipment damage and disruption in operation in an industrial facility due to fire

1. In an industrial facility, two boilers are turned on to produce hot water needed for cleaning equipment at the start of daily operation.
2. Leftover packaging materials were left stacked near one of the boiler near an exit door the prior night. These should have been placed into a recycling bin very early in the morning, but for some reason, this was not the case for this particular day.
3. The stack of leftover packaging materials tipped over and some of the materials landed very close to the boiler's heating element
4. Material near the heating element ignited
5. Fire spread to the entire stack of leftover packaging materials
6. Smoke accumulates in the enclosed space where the two boilers are located
7. Any person within this enclosed space may suffer from severe smoke inhalation or burn
8. The boilers themselves may catch fire or the smoke might damage them, particularly the external components that are made of plastics, rubber, and glass making both boilers inoperable
9. Fire may spread to other combustible materials nearby such as boxes of recent delivery receipts and inventory of packaging materials
10. Fire may spread to outside the enclosed space where the boilers are located and further damage more equipment and possibly result in more injuries

As an example, smoke detectors, whether in a household or in an industrial facility, have the primary purpose of treating the target risk event of property loss, injury, or death due to fire. A smoke detector being able to detect the presence of smoke provides opportunities to completely stop or slow down the causal chain of events that may lead to the target risk event by allowing evacuation and suppressing a fire. From Table 4.1, designating and clearly marking of floor spaces near the boiler as a material-free zone may serve as a visual identification of any hazardous materials near the heating element; for example, detection of combustible materials near the boiler may provide opportunity to prevent segment 2 to progress further to segment 3. Furthermore, the detection of smoke before the spread to the entire stack of leftover packaging materials may have provided the opportunity to prevent segment 4 to progress further to segment 5.

4.2.2 CONTROL

Once the causal chain of events or the target risk events are detected, the opportunity is now presented to control them if at all possible. Control may be in the form of slowing down or totally stopping the progression of the causal chain of events. Some common forms of risk control strategies are separation and suppression.

Separation pertains to the practice of increasing the spatial distance or providing barriers between hazards and the remainder of the system. In a household or an industrial setting, evacuation is a form of separation by providing adequate distance between the fire and humans (and other valuable movable assets). In an industrial setting, such as that described in Table 4.1, the adequate distance between combustible materials and the boiler could have been implemented by having a material-free zone around the boiler and could have prevented progression of the causal chain of events. The evacuation of personnel after the smoke has been detected could have also provided separation between the personnel and the hazardous smoke and heat and could have prevented injury due to smoke inhalation (i.e., segment 7). Furthermore, fire-resistant doors serving as barriers could have provided separation as part of passive fire protection, possibly controlling the spread of fire outside the enclosed area.

Suppression pertains to the intervention to lessen the magnitude or totally eliminate hazards that cause the target risk event. In the incident of a fire, suppression of fire using various forms of fire extinguishing mechanism may be used to lessen or eliminate the presence of fire. For the scenario described in Table 4.1, a fire suppression system could have been used at some point in time after a smoke or fire has been detected, which could have prevented further damage to equipment.

4.2.3 RECOVERY PLANNING

Recovery planning pertains to the development of a plan of actions in anticipation of the target risk events to primarily assure the continuity of operation, as well as to reduce further accrual of undesirable consequences. These plans are also known as emergency response plans, disaster recovery plans, and may itself include detection and control, as well as risk transfer. Parts of these plans may require actions both before and after the occurrence of risk events.

Duplication pertains to having at least one spare or backup element that can perform the same or similar functions as a primary element that is deemed to be critical to the operation of the system. These spare and backup elements are intended to perform the function of the primary element in case the latter cannot function due to the target risk events. As an example pertaining to segment 9 in Table 4.1, a duplicate of recent delivery receipts in another area would have provided the function of the original receipts in the continued settling of financial accounts after the fire. Otherwise, the settling of financial accounts may be disrupted while information on the receipts are recovered.

Dispersion pertains to the separation of elements of the system performing similar functions while both of them are being utilized. Dispersion is distinct from duplication because in dispersion, elements are not spare or backup but are rather all being utilized at the same time. As an example pertaining to segment 8 in Table 4.1, having the two boilers in two different locations could have saved one of the boiler from being damaged by the fire and may contribute toward continuous operation of the industrial facility even if not in full capacity.

Transfer (including sharing) of undesirable consequences implies that the consequences of a risk event are either completely or partially reassigned to another entity or system. This reassignment of the consequences is usually based on the agreement between two distinct systems rather than an actual physical reassignment. Typical risk transfer or sharing agreements are those involved with the purchase of insurance products and services. For the example illustrated in Table 4.1, fire insurance could help towards repair or replacement of damaged boilers, other equipment and facility, disrupted operations, and alleviate any longer-term financial costs from the treatment of injured personnel.

4.3 FUNDAMENTAL STEPS IN RISK TREATMENT

For convenience, the 1 + 6 general guiding questions in risk management (RM) in Table 2.1 are repeated here.

0th What should go right?
1st What can go wrong?
2nd What are the causes and consequences?
3rd What is the likelihood of occurrence?
4th What can be done to detect, control, and manage them?
5th What are the alternatives?
6th What are the effects beyond this particular time?

As described in Chapter 2, the 1st, 2nd, and 3rd questions are those usually associated with risk assessment process culminating in some form of analysis, not just about particular target risk event but also a comparison among other risk events.[2] On the other hand, the 4th, 5th, and 6th questions are those usually associated with RM wherein actions on how to use resources for affecting risks are rationalized and justified.[3] To a great degree, the steps in designing plans on how to treat risks follows these 1 + 6 questions in parallel and can be summarized in the following four steps.

1. Identification of treatment options
2. Development of an action plan
3. Approval of an action plan
4. Implementation of the action plan

4.3.1 IDENTIFICATION OF TREATMENT OPTIONS

In the first step—identification of treatment options—we need to choose which risk events will be targeted based on their importance and identify appropriate detection and control strategies based on the causal chain of events. The following guiding questions may facilitate identification of treatment options.

Which risk event(s) needs to be treated? After obtaining answers to the 1st, 2nd, and 3rd guiding questions in RM, the risk analysts can now compare the relative importance of various risk events and effectively establish prioritization among these risks as needing treatment. The prioritization may be based on the combination of the chances of occurrence and degree of consequence. Oftentimes, a risk matrix similar to that in Figure 4.1 is used to compare among risk events and facilitate their ranking or prioritization. A risk event with a high chance of occurrence and high degree of consequence may be deemed of high priority, such as those that may fall in Zone A of Figure 4.1. Risk events with lower chances of occurrence and lower degree of consequence may be of the lowest priority and may fall in Zones B and C. Using a risk matrix similar to that in Figure 4.1 may be appropriate in cases where the consequences of risk events can be expressed in similar measures, such as a monetary unit. However, there may be cases where expressing consequences in the same measure may not be possible or acceptable, such as in the case of some risk events resulting in injury and some resulting in equipment damage. Another way to compare and prioritize risk events may be based on categories rather than a matrix such as those shown in Table 4.2. Categorization

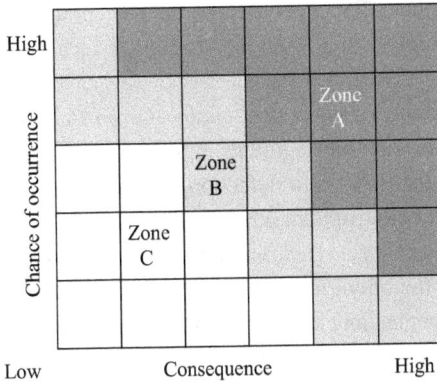

Figure 4.1. Risk matrix showing possible way to compare and prioritize risk events.

Table 4.2. Categorization of risk events based on consequence to facilitate comparison and prioritization

Consequence if risk event occurs	Chance of occurrence	Category
Injury or loss of lives	Low, moderate, or high	High priority
Financial cost of more than $1M	Moderate or high	High priority
	Low	Medium priority
Disruption in operation lasting more than 24 hours	High	High priority
	Low to moderate	Medium priority
Irreversible environmental damage	Low, moderate, or high	High priority
Reversible environmental damage	Moderate or high	Medium priority
	Low	Low priority

allows the analyst to distinguish among types of consequences based on criteria relevant to a particular system without the need to express them using the same measure. For example, a system that greatly values the environment may have a distinct criterion for risk events that may result in environmental damage. Nonetheless, such categorization of risk events would still be based on some assessment of the chance of occurrence coupled with the consequence if they occur.

Which segments in the causal chain of events can be detected? Certain segments in the causal chain of events may be more amenable to detection than other segments based on the mechanics of the phenomena, available technology for detections, and other factors (e.g., ethical and legal factors).

Which segments of the causal chain of events can be controlled and how? In the same way that segments in the causal chain of events may be more amenable than others to detection, some of these segments may be more amenable to control than others.

If the risk event happens, how can the system recover quickly? Important functions in the operation of the system are identified for the purpose of analyzing whether any of them can be duplicated or dispersed. This applies not only to equipment and information as illustrated in Table 4.1 but also to personnel through cross-trainings. Finally, options to purchase insurance products and services also need to be considered based on economic and legal factors.

4.3.2 DEVELOPMENT OF ACTION PLAN

In the second step—development of an action plan—we need to generate combinations of particular risk treatments that will be effective for a particular system and target risk event and the accountable party.

Which risk treatment strategies will work well together? Risk treatment strategies are not mutually exclusive and effective action plans that are usually made up of combination of strategies, albeit in various degrees.

Who will be accountable for implementing the plan? Upon determination of risk treatment strategies, the parties (individuals, job descriptions, teams, or organizations) that will be most reasonably held accountable for their implementation are identified.

4.3.3 APPROVAL OF THE ACTION PLAN

In the third step—approval of the action plan—we need to obtain the approval of the action plan from the management and from responsible parties.

How much will this cost? From the management's perspective, there will most likely be a need to economically rationalize the choice of risk treatment strategies based on the benefit that the system will get from such plan compared to the cost of implementation.

Who will lead the implementation? Management commitment is also needed to coordinate among the various responsible parties and assure that required resources for implementation are made available. From the perspective of the parties responsible for the various treatment strategies, their commitment based on motivation and incentives are needed for long-term effectiveness of the plan. A single party—individual or team—is usually assigned the accountability of assuring the plan is implemented. This party responsible for the overall plan is usually labeled as the risk manager, safety officer, or other similar job title.

4.3.4 IMPLEMENTATION OF THE ACTION PLAN

In the fourth and final step—implementation of the action plan—usually falls under the purview of the party responsible for the overall plan, for example, safety officer.

How can the plan be improved? Since this is a long-term responsibility, part of the implementation is the continuous improvement of the plan based on experience of which part of the plan works and which do not, based on changes internal to the system, for example, new product line and realization of new target risk events, as well as changes external to the system, for example, evolving detection and control technologies, new insurance products and services, and legal and regulatory changes.

What are the residual and emerging risks? We need to recognize that there will be no plan that can totally eliminate the target risk events, rather the plan simply reduces their likelihood or consequence to acceptable levels—these are termed as residual risks. But due to the dynamic nature of systems, these levels need to be continually monitored and the plan must be periodically revised to keep the residual risks within acceptable level. This monitoring and periodic evaluation of the plan also enables the identification of emerging risk events—those that may have not been previously identified, or that were identified but were not deemed worth any treatment.

4.4 RISK TREATMENT AND RISK RETENTION OR ACCEPTANCE

An important aspect of risk treatment is the recognition that there will be more risk events than there are resources to treat them all. Even for those risk events that are subjected to treatment, there will always be some residual risks, albeit at acceptable levels. Hence, part of the bigger picture of operational RM is the realization that it is only practical to expect some degree of risks that will have to be retained and accepted even after treatment. Recall from Chapter 3 the philosophy of as low as reasonably practical (ALARP) holds that the residual risk shall be as low as reasonably practicable. As zero risk cannot be attained and the cost of addressing risk can be excessive if pursued without limits, ALARP provides an approach to help derive justifiable and tolerable risk levels. The approach involves judgment in weighing the risk and the benefit. Thus, the concept also includes accepting a certain level of risk because of the deliberate decision to bear it rather than address it. The ALARP principle is also similar to the SFAIRP (so far as is reasonably practical) principle.

4.5 SYSTEMS PERSPECTIVE ON TREATING RISKS

In Chapter 1, the importance of systems approach to managing operational risks was discussed. Here are some important notes on risk treatment based on the recognition that systems are collections of interrelated elements contributing toward a common goal and the environment wherein these systems are parts of.

4.5.1 MULTICAUSAL NATURE OF RISK EVENTS

Table 4.1 provides an illustrative example of a causal chain of events leading to a particular target risk event. Even in a fairly simple real scenario, causal events may not be linear, there could be multiple ways for the target risk to occur, and some of these causes may not be easily evident.

4.5.2 SYNERGISTIC EFFECTS OF RISK TREATMENT STRATEGIES

Aside from reducing the target risks, the same set of mitigation strategies may also have the synergistic benefit of reducing other none-targeted risks whether intentional or not. As an example from segment 2 from Table 4.1, keeping floor space near the boiler free from any material may help prevent fire and may also prevent accidents as personnel try to use the exit door near that area.

4.5.3 CREATION OF COUNTERVAILING RISKS

In contrast to having a synergistic effect, some risk treatment strategies may result in increasing other events than the target risk events. As an example from segment 9 from Table 4.1, having duplication of receipts in electronic format may also create risks of confidential data pilferage due to cyber hacking.

4.5.4 RATIONALIZING COST OF TREATMENT

Similar to the traditional cost estimation and investment accounting, the cost of implementing risk treatment plans needs to be rationalized by comparing the benefits and the costs. However, risk events are inherently uncertain and do not lend themselves to the traditional benefit–cost analysis. The common approach is to classify the various cost of the strategies based on (a) cost of prevention meant to reduce the chances of occurrence of risk event, (b) cost of reducing the consequence if the risk event occurs, and (c) the residual consequences of the risk event.

4.5.5 BENEFITS OF RISK AVOIDANCE

A well-rationalized risk treatment plan would be one in which the benefit of avoiding some of the direct consequences of risk events are more than the sum of the cost of prevention and the cost of reduction in the consequence if the risk events occur. However, estimating these benefits to allow comparison with the costs can be challenging. From a systems perspective, this challenge may be traced to the variety of objectives of various elements of the system, as well as the lack of a commonly agreeable way to express benefits. As an example, the traditional cost analysis may estimate cost in terms of monetary unit, while the treatment of system safety risk events may have the benefits of avoided injuries or lives saved. The notion of the value of a hypothetical or statistical life may facilitate such a comparison of cost and benefits. However, the success of using pseudo measures such as the value of statistical life is predicated on its acceptance by the various stakeholders and decision makers within and outside a system trying to implement a risk treatment plan.

4.5.6 ECONOMICS OF RISK TRANSFERS

Why would someone assume the risk you are not willing to assume yourself? The underlying assumption in transferring risk is that another system, for example, insurance companies (also known as the insurer) will be willing to assume some if not all the financial burden resulting from a particular risk event in return for a predetermined amount of money such as insurance premiums from the original risk holder (the insured). To put it simply, this transfer of risk will happen if the insurer has the ability to pool or bring together enough of this money from many insured such that the expected payout if the risk event occurs is significantly less than the sum of the collected insurance premiums. This pooling of risks enables the insurer to operate at a profit and still satisfy its obligation for payouts to the insured. This is common for risk events with high consequences but low chance of occurrence. Common examples of such insurable risks are health insurance and drivers' accident insurance in which there is enough number of those willing to pay the premium and the chance that everyone will need a payout all at the same time is very low. From a systems perspective, this highlights the wide range of environments in which the insured systems are operating.

4.6 SUMMARY

Risk treatment is a general term to indicate modifying a system or its elements for the purpose of affecting particular risk events deemed important based on the chances of occurrence, the consequences if the risk events happen, and other factors. The general objectives in risk treatments are (a) reduction in the chances of occurrence of the risk events and (b) reduction in the consequences if the risk events occur. The fundamental steps and corresponding guide questions in developing a plan for risk treatment are as follows.

1. Identification of treatment options
   - Which risk event needs to be treated?
   - Which segments in the causal chain of events can be detected?
   - Which segments of the causal chain of events can be controlled and how?
   - If the risk event happens, how can the system recover quickly?
2. Development of action plan
   - Which risk treatment strategies will work well together?
   - Who will be accountable for implementing the plan?
3. Approval of action plan
   - How much will this cost?
   - Who will lead the implementation?
4. Implementation of action plan
   - How can the plan be improved?
   - What are the residual and emerging risks?

Finally, the systems approach needs to be considered throughout the development and implementation of risk treatment plans. Consideration needs to be placed on possible challenges brought about by the multi-causal nature of risk events, synergistic effects of risk treatment strategies, resulting countervailing risks, the need to rationalize cost of treatment, estimating the benefits of risk avoidance, and economics that may allow or prevent risk transfers.

CHAPTER 5

Risk Monitoring, Reporting, and Review

5.1 INTRODUCTION

After identifying and assessing the risk and selecting risk treatment response plans, the next important function is risk monitoring, reporting, and review, as shown in Figure 5.1. Here the purpose is to monitor the implementation of the risk response plans and actions and evaluate the effectiveness of the current risk management processes. A key part of this function is assessing how the present risk management plans and processes can be improved to respond to the current and emerging environments.

5.2 RISK MONITORING OBJECTIVES

1. To determine whether the particular operational risks (and their causal chain of events) are under control or not. This includes tracking identified risks and checking that significant risks remain within acceptable risk levels.
2. To assess the effectiveness of the risk treatment, response plans, and actions in addressing the identified risks. This includes gaining relevant information during the monitoring and review process to further assess, control, and eliminate critical risk events.
3. To evaluate the robustness of risk controls over time as the system evolves.[1] This also includes monitoring residual risks, identifying gaps and new risks, as well as retiring invalid or outdated risks.
4. To keep track of the evolving needs and requirements of the systems. Remember that new needs and requirements are often times accompanied by new risks that need to be controlled.

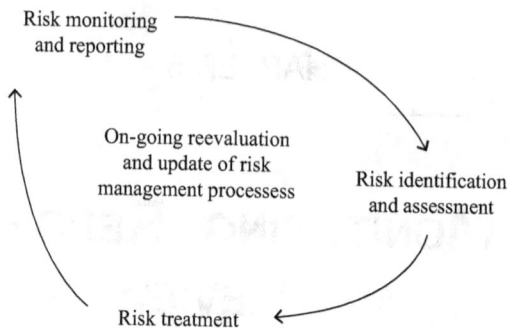

Figure 5.1. Risk management functions.

Monitoring and management of risks is one of the main elements in the risk treatment and risk management process.[2] There is no one specific methodology or process that can be applied to monitor operational risks in all systems. For example, monitoring operational risks in a bank is different from monitoring operational risks in a production system. Based on a survey of the literature and case studies, the following common themes help to conduct an effective monitoring process.

- Performance monitoring requires both an understanding of the context of the risk event and the methodology used for the performance monitoring as well as an appreciation of the uniqueness of the operational risk of the system. If the methodology used is not appropriate for the system, there will likely be unintended consequences and poor operational risk management and performance.
- For a reliable monitoring process, comprehensive treatment and assessment of risk events are necessary. It is essential to have a clear set of definitions of the operational risk under monitor. This includes identifying the appropriate risk indicators.
- The development of flexible methodology to accommodate any adjustment or changes in the system. It is naive to consider that the operation environment is always static, and therefore, operational risk can be controlled or avoided in the systems all at once. With increasing consumer demand and expectations resulting from advancing technologies, many systems are dealing with shifting dynamic environments. Therefore, it is necessary to adopt a flexible methodology to mitigate and control operational risk to the minimum.

- Full appreciation of the surrounding environments where systems operate. It is essential to understand the nature of the operations' environment. As will be discussed through Chapters 6 and 7, there are two types of environments: static and dynamic. In the static environment, the system is bounded apart from the environment, which means that there is no or minimal interaction between the system and the environment. In the dynamic environment, there is continuous interaction between the system and the environment. While it is relatively easy to monitor the operational risk in a static environment, it is a much more difficult task in a turbulent shifting environment.

5.3 THE BASELINE: RISK MANAGEMENT PLAN

The culmination of the risk processes discussed in Chapters 1 to 4 is a working document often called the *risk management plan*. This report typically becomes the baseline for any risk monitoring process. In general, the report should include the following parts.

1. Description of the scope of the risk management plan (refer to Chapter 2)
2. Risk identification and assessment (refer to Chapter 3)
3. Risk treatment, response, and action plan (refer to Chapter 4)

The risk management plan is essentially a compilation of the broader system's risk profile or risk portfolio. Note that the report should cover the 6 + 1 questions (Table 2.1) as discussed in Chapter 2. It defines the scope of the risks being tracked and quantifies or measures the risks in terms of their estimated likelihood and impacts. It should further describe established acceptable risk level (i.e., As Low as Reasonably Practicable [ALARP] from Chapter 3) and detail the treatment response plans and actions for each identified risks. Included in the plan are the processes established for detecting, tracking, and documenting risks, as well as persons responsible for each risk response and actions. The forms discussed in Chapter 3 can be used to document the particulars of the risk management plan for each part, subsystem, or process.

The report can also include the risk matrix discussed in Chapters 2 and 4. It offers a good overview and summary of the comparative risks. The summary of the risk management plan should include a list of risks requiring immediate response, a list of risks requiring response in the near term, a list of risks for additional analysis and response, and a watch list of low-priority risk. The report should also highlight any developments and trends from the risk data and risk assessment including an analysis of new and emerging risks.

5.4 TRACKING: RISK DATABASE

The most common database compiled for risk is what is known as the risk register or risk log. The database and information requirements are extracted from the risk management plan. It often replicates or digitizes the tabular forms in Chapter 3. Though, more importantly, the risk register *tracks the actual occurrences* of the risk events. Further, the risk register can document conditions leading up to the risk event, the response and actions taken, as well as the actual outcomes. The specifics are often included in an *incident report*. An incident report or accident report is a form that is filled out in order to record the details of the risk event.

Other types of risks such as cost risk, schedule risk, programmatic risk, and technical risk have certain indicators that may be used to monitor status for signs of risk. Tracking the progress of key system parameters can be an indicator of risk or key risk indicator (KRI). Cost and schedule risk, for example, are monitored using the outputs of the cost–schedule control system or some equivalent technique. Thresholds for KRIs can be defined to initiate a risk treatment response and action when thresholds are breached. Examples of this are project cost overrun and project schedule slippage.

The risk register when translated into a digital database or information architecture offers several enhanced functions. A database can handle the recording and compilation of detailed historical risk information. Also, a database can enable automated tracking and signaling. It can be used to identify risks that are within acceptable levels, risks that are out of control, and risks that are trending toward out of control. These KRIs can be captured and stored as digital data to be used for future planning and decision making.

Further, the risk database should track the implementation of treatment and response plans, plans that are implemented, plans that are not implemented or partially implemented (including schedules and milestones), as well as their costs and effectiveness in addressing the risks (including any instances of overcontrolled risks). Figure 5.2 shows a screenshot example of a simple risk database.[3]

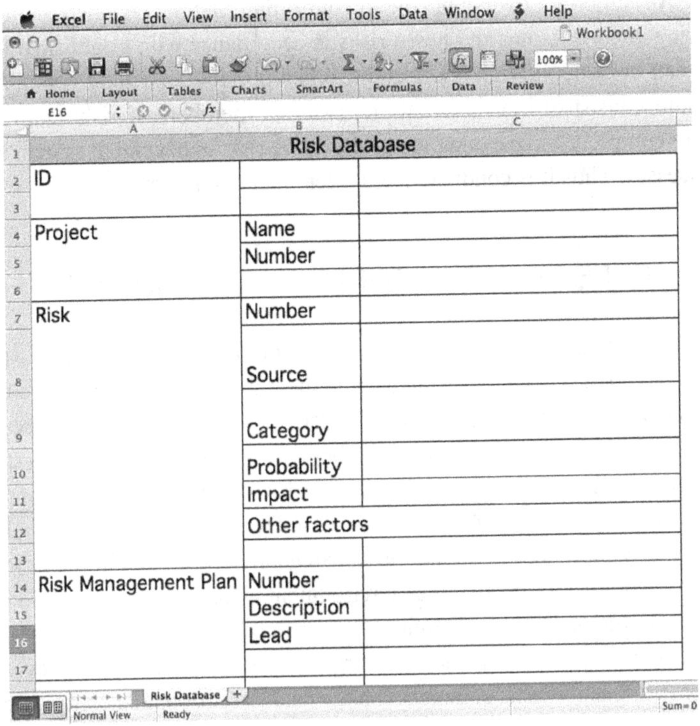

Figure 5.2. Screenshot example of risk database.

5.5 RISK REVIEW AND AUDIT

The risk review examines both intended and documented results of the risk response and action plans in dealing with identified risks. Included in the risk review is an evaluation of the residual risk against established risk levels. Residual risk assessment considers both the risks as previously identified and the resulting risk from the response mechanisms and the control activities implemented. The process evaluates the effectiveness, cost, and efficiency of the activities and its adequacy in providing reasonable assurance that the likelihood and impacts of the risk events are brought down to the established acceptable risk level (i.e., ALARP).

It is critical that the risk review evaluates the significant risks that fall outside of the acceptable risk levels. This may happen due to changing systems and policies or changing conditions and environments. These risks should be addressed with appropriate action plans to bring the risk within established levels. Risk treatment and response plans should be updated accordingly.

An important oversight role in confirming that management is monitoring and managing all known risks in accordance with established levels can be done through a risk audit. When possible, the audit committee should consist of both internal and external members. A risk audit is a more comprehensive evaluation of the entire risk management system and processes. Thus it is conducted less often than risk reviews.

5.6 EFFECTIVE PERFORMANCE MONITORING

To be effective, it is emphasized that the performance monitoring processes should be treated as an integrated component in the system. There are two key elements in achieving an effective risk monitoring and review process. First, clear *communication channels* between the members of the system increase the level of awareness with regards to the implementation of risk treatment and controls. All stakeholders in the system must know specifically why particular risk controls or safeguards have been implemented. Clear communication creates an environment where every member knows precisely his or her responsibilities toward addressing the risk. Communication is important especially in large systems that have a high level of interaction, ambiguity, emergence, huge data collection, and complexity.

Second, it is necessary to establish a *flexible* monitoring process. The design of risk monitoring procedures should be flexible so that they can adapt and respond in a cost-effective manner to threats arising from emergence, turbulent environments, uncertainty, and contextual issues of a dynamic nature. Flexibility is the ability to add, adjust, or remove both physical components and functions. The level of flexibility should not cause the monitoring process to lose its identity; rather, it should provide an environment of trust where members of the system can share their proposed changes and strategies.

5.7 SUMMARY

Performance monitoring should be implemented to control and manage risks at established levels and to ensure the appropriateness and effectiveness of the risk treatment and response over time. These are done by using information from sources such as risk management plan and risk databases.

CHAPTER 6

Systems Thinking and the Risk Manager

6.1 INTRODUCTION

Complex systems can be described as a large number of entities that interact to produce behaviors or performances that cannot be understood simply as a function of the individual entities. This is where the commonplace phrase "the whole is greater than the sum of the parts" comes from, dating back to Aristotle. Managing risks in these systems has proven to be a difficult endeavor.[1] It is considered relatively easy to optimize safety within a simple system where there are a small number of entities with clear cause–effect relationships, but it is a much more difficult task to produce consistent safety performance within a complex system. Risk managers break problems into parts, study each part in isolation, and find the appropriate solutions to deal with each problem. Although, in theory, this appears to be an effective method to deal with problems and make complex tasks and subjects more manageable, in reality, it has many limitations when applied to complex "real-life" situations. According to Senge, breaking problems into discrete manageable elements to solve is an insufficient concept when applied to "real-life" situations.[2] Instead, system behavior in real-life situations comes from interactions among parts and cannot be determined by the isolated functions of individual parts.

6.2 THE DIFFICULTY WITH COMPLEXITY

Dealing with complexity and its associated problems is a reality for engineers dealing with today's complex problems, especially those in managing operational risks. The problems and behaviors associated with increasing complexity continue to confound our capabilities to deal with modern systems. Many large-scale organizations have integrated their existing operations, functions, and systems to work together to achieve higher-level goals that are beyond the capabilities of individual systems and have increased the interdependence of interactions and reduced the autonomy of single systems.[3] This drive for increased integration increases the difficulty for operational risk managers in engineering and managing risks in complex systems. Thus a primary challenge for operational risk managers is to determine the risks stemming from the interaction and integration inherent in complex systems and to deploy appropriate tools for better risk control behaviors and procedures within these systems.

Risk managers are starting to recognize the challenges associated with complexity and to address the challenges being invoked by the realities of the complex systems that they are charged to manage.

According to Jablonowski,

> ... today's businesses face exposure to accidental loss from a variety of perils, both natural and man-made. These include fire, windstorm, earthquake, legal liability for defective products and the hazards of the workplace. Dealing with these exposures is the job of the organization's risk manager.[4]

6.3   RISK AND COMPLEX SYSTEMS

Attempting to manage risks in complex systems requires knowledge not only of technology but also of the inherent human, social, organizational, managerial, and political and policy dimensions. In effect, a holistic perspective integral to systems thinking, as discussed in Chapter 1, is necessary for risk managers to effectively design and integrate risk management plans into complex systems. Our intent is to shift focus to supporting the necessity of having a cadre of risk managers who can effectively understand the nature of complex systems.

One could ask how many risk events could be attributed to misunderstanding the nature of complex situations. Although the answer is not definitively known, we might anecdotally surmise that there are many mistakes that might be attributed to a lack of understanding and accounting for the complexities in systems.

6.4 SIMPLE SYSTEMS VERSUS COMPLEX SYSTEMS[5]

The selection and utilization of effective risk management tools discussed in Chapter 3 requires that risk managers appreciate the uniqueness of the problem (simple versus complex), context (conditions and factors that impact the solution and the deployment of the solution), and the methodology (the particular approach to deal with the problem). It is important to match particular tools to the specific application because a mismatch is not likely to produce the desired outcomes to efficiently and effectively manage risk events. Not appreciating the nature of the system (simple versus complex) may come with substantial costs in accidents and human injury or death because managing simple systems is different from managing complex systems.[6] For conciseness, a set of differences between simple and complex systems most addressed in the literature have been identified in Table 6.1.

- This list is certainly not exhaustive but is intended to show that simple systems are significantly different from complex systems. The implication for risk managers is that any misunderstanding or diminishing of the importance of these differences can lead to one or a combination of the following problems.
  a. Falling into a type III error,* which is solving the wrong problem precisely and in the most efficient way.[7]
  b. Ineffective resource utilization, resulting from insufficient resources for addressing complex systems when they are misclassified as simple systems. In contrast, wasting resources by treating a simple system as complex. Resources span the spectrum of money, time, machines, physical effort, and others related to the exchange of information, energy, or materials.
  c. Selecting risk management tools and techniques that are incompatible with the nature of the problem (simple versus complex) and the relevant context.
  d. Formulating the problem incorrectly in ways that could jeopardize the integration of system risk management plan. Formulating and understanding the problem (risk identification) is the first step in risk management as previously discussed in Chapter 2.

---

*Type I error: Rejecting the null hypothesis H_0 when it is true. Type II error: Failing to reject the null hypothesis H_0 when it is false.

Table 6.1. Simple system versus complex system

Area of focus	Simple	Complex
Relationship between parts	The relationship between the interacting parts is clear and easily defined (linear relationship)	The relationship between the interacting parts is ambiguous (nonlinear relationship)
Cause–effect	Cause and effect are obvious and direct	Cause and effect are very difficult to associate
Boundary (formulation of the problem)	Well-defined and understood boundary	Ambiguous boundary
Focus and treat the problem	Focus on the parts of the system	Focus on the interaction between these parts
Number of parts	Small number of entities	Large number of entities
Output of the system	Predict deterministic output (consistency)	Probabilistic
Knowledge	Not complete but adequate to address the problem	Not complete
Predetermined attributes	Easy to know many things in advance before starting analyzing the risk(s)	Hard to know things in advance or remain constant
Interaction organization	Organized and apparent structure	Fluctuating structure and relations between parts
Nature of the system	Generally static	Dynamic: Many forces can change the system over time such as involving a human in the system
Classification the system according to the environment effect	Closed system has no or minimal interaction with the surrounding environment	Complex system is open to influences from the environment
Problem area	Well defined and understood	Emergent and difficult to understand
Affected by behavior influences	Not subjected to behavior influence	Easily affected. Actually, complexity can be a function of observer

To avoid these problems, risk managers must understand the nature of the system they are dealing with before selecting the risk management techniques to identify and mitigate the risks. In fact, the more complex the system is, the greater is the risk involved. The question becomes *what are the systems thinking characteristics risk managers need to understand complex systems problems?*

6.5 ATTRIBUTES OF COMPLEX SYSTEMS

While the previous section showed the differences between simple systems and complex systems, the legitimate question now becomes what are the main attributes that constitute a complex system. Throughout the history of complex systems there have been multiple characteristics, articulations, and definitions. This section focuses on the main attributes that emerged most frequently (from the coding of the literature sources).[8] These attributes, shown in Figure 6.1, are the most dominant in the complex problem domain.

The complex problem domain faced by risk managers is marked by a dramatic increase in information and technology. This frequently requires risk managers to face challenges of making decisions at various levels of their system amidst near-overwhelming complexity. Based on the coding analysis, the seven main attributes of the environment of the current complex system encountered by risk managers are presented as follows:

1. Contextual Issues: Complexity exists when a situation exhibits a high level of interrelationships among the elements and their parts, a high level of uncertainty and ambiguity, an emergence, a large amount and flow of data, an incomplete knowledge, and a highly

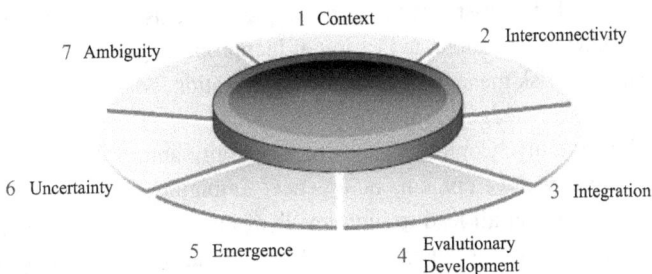

Figure 6.1. Attributes of the complex system.

dynamic nature. Complexity also entails contextual issues—specific external influences, characteristics, or conditions that influence and constrain the deployment of solution(s). Contextual issues include the political, managerial, social, and cultural considerations as well as the resources or funding, organizational, policy, and technical dimensions.

2. Interconnectivity: Interconnectivity in the complex problem domain includes a high level of interrelationships among the individual systems and their components, human interaction (social–technical problems), and the interactions of the systems' components (hardware and software).

3. Integration: To deal with the dynamic environment, many organizations tend to "bring together" their existing or external systems to work together to produce new capabilities or behaviors that cannot be achieved by the individual systems. This integration of multiple systems, which is referred to as system of systems, makes systems more complex and dynamic and can include people, information, technology, hardware and software, and many other elements.

4. Evolutionary Development: The environment within which complex systems operate is highly uncertain and dynamic, with potentially rapid shifts occurring over time. In addition, multiple perspectives, stakeholders and shareholders who have a direct or an indirect impact on the system contribute to the complexity and divergence of such systems. This leads to issues such as the evolution of needs (requirements) over time and the necessary reallocation of scarce resources based on these shifts.

5. Emergence: Emergence is recognized as unintended behaviors or patterns that cannot be anticipated. These unforeseen behaviors result from the integration of systems and the dynamic interaction between the individual systems and their parts, the surrounding turbulent environment, and the high levels of uncertainty and ambiguity characteristic of modern complex systems. These unintended behaviors are not an indication of inappropriate tools or techniques used in risk management, or their application, but rather an insufficient knowledge about complex systems.[9]

6. Uncertainty: Complexity, interconnectivity among the parts, evolutionary development, emergence, ambiguity, and the dynamic environment all lead to high levels of uncertainty in complex systems. This uncertain nature negatively impacts our knowledge level of a system, makes decisions tenuous, and limits the applicability

of more traditional system-based approaches such as deterministic (e.g., algebraic) and probabilistic (e.g., statistics) methods of analysis.

7. Ambiguity: Ambiguity is the lack of clarity concerning the interpretation of a system's behavior and boundaries. Boundaries in complex systems are arbitrary, incomplete, permeable, and subject to conceptual variability. Unclear boundary definition and management casts doubt on the primary boundary questions that must be answered: *What is included and excluded for the purposes of analysis? And what is the criteria for inclusion or exclusion within the boundary?* Boundaries can and do change over time based on our evolving understanding and knowledge of the system. In technical terms, anything outside the boundary of the system is part of the surrounding environment and anything inside the boundary is part of the system.

Because the domain of complex systems is characterized by these seven attributes, the improvement of managing risks in complex systems remains a dilemma for risk managers. Within the complex system domain, these professionals must still design and maintain a flexible risk management system that minimizes the consequences of risk events.

Although there are many techniques and tools that can be used in risk analysis, they appear to lack the perspective of viewing the problem domain from a holistic standpoint. In order for risk managers to make suitable decisions regarding assessed risk(s) in complex systems, they need to explore and articulate the problem as a whole—taking a truly systems view. The selection of the tools and techniques to analyze risks are based entirely on the formulation of the problem. Proactive risk management occurs when risk managers forecast risk potential and then adopt systems management activities that control or mitigate that risk potential.[10]

In addition to proactive risk management, risk managers can enhance effectiveness by using systems thinking, based in the foundations of systems theory and principles, to achieve a more holistic risk identification. This level of "systemic thinking" can enhance understanding of the interactions among systems' element and context to more rigorously and "holistically" understand the nature of the problem domain. Even though systems thinking does not provide a universal solution, it does offer a high level of holistic thinking to deal with the attributes of complex system landscapes as described earlier.

6.6 THEMES IN SYSTEMS THINKING

Systems thinking is not a new domain; it has been around for decades. In fact the earliest roots of systems thinking can be traced back to Aristotle who introduced the idea of holism as a main characteristic of systems thinking. As described in Chapter 1, systems thinking is the thought process that develops an individual's ability to speak and think in a new holistic language and is important in understanding and treating problems as a whole unit.[11] Throughout the history of systems thinking many definitions have been emerged and one could find that (1) there is no one widely accepted or generalized agreement on the definition of systems thinking and (2) systems thinking is not specific to one domain or another, but rather it spans several domains including social, managerial, biological, and many others.

In systems thinking, it is important to understand the principles and laws of systems theory that capture general principles applicable to all systems regardless of their domain of existence.[12] Although discussing each principle and law of systems theory is beyond the scope of this chapter, there are several themes in systems thinking that are particularly relevant in managing operational risks.[13]

- Holism versus reductionism: In large complex problems, the focus should be on the whole instead of one particular thing. System performance or behavior is generated from the interaction of the elements, not from individual elements.
- Turbulent environment versus static environment: In dynamic environments, it is important to appreciate the effect of the environment on the system both negatively and positively. A system is influenced by the externally driven impacts (perturbations) from the environment.
- Optimizing versus "satisficing:" In large complex problems, it is difficult to achieve one best solution. There are always multiple working (good enough) solutions that need to be considered.
- Nontechnical problems (sociotechnical) versus technical problems: Complex systems have a combination of both technical (technology) and nontechnical aspects (culture, human or social, policy, politics, power, etc.). There is a need to have the skills to formulate the problem domain holistically, including both the technical as well as nontechnical perspectives.
- Interconnectivity and interrelationships: Understanding the structure and the importance of the relationships of the system's parts by

providing a systemic perspective. A system is defined not only by the elements it includes but more importantly by their interrelationships and products of their interactions.

6.7 BENEFITS OF A SYSTEMS APPROACH TO RISK MANAGERS

As discussed earlier, systems thinking provides a significant contribution to helping risk managers understand the complex system problems they face from a holistic perspective. The best procedure is to detect all risks in the system and then minimize them to a level of insignificance (or acceptance). However, the attributes of a complex system landscape (Figure 6.1) make it especially difficult for risk managers to identify and eliminate all risks, particularly if we accept the notions of emergence, ambiguity, and incomplete understanding, which was addressed in Chapter 4.

One of the main responsibilities of risk managers is to select an existing risk management tool that is compatible with the nature of the complex problem and the context. This is not an easy task in systems where complexity, ambiguity, and uncertainty are always present. Hammitt mentioned that to determine "whether one tool is better than another, it is necessary to evaluate whether the harms imposed on some people (e.g., costs of compliance) are offset by the benefits conferred on others (e.g., reduced health risk)."[14]

It should be emphasized that in order to address the attributes of complex systems (Table 6.1), risk managers must shift to a paradigm with a more systemic (holistic) perspective. We suggest that this paradigm, based in systems thinking and systems theory, is more consistent with the problem domains and complex systems characteristic of those faced by risk managers. The following are some of the implications of systems thinking for risk managers with respect to the attributes of complex systems.

Understanding the big picture. By applying systems thinking, risk managers can better and more rigorously understand the big picture of the problem by looking at the problem as a whole. Understanding the problem as a whole unit is necessary to formulate the true problem and correctly identify the potential risks. This is important to avoid managerial errors such as type III error, solving the wrong problem precisely, resulting from *selecting the wrong stakeholders, setting too narrow a set of options, establishing boundaries that are too narrow, incorrectly phrasing the problem, and failing to think systemically.*[15]

Understanding connectivity and the causal chain of events. Employing a holistic perspective can be instructive in understanding the interrelationships between components within a complex system. Some have used the term "messes" to describe the high level of interconnectivity that produces "sets of problems" within complex systems. Hazard identification processes require risk managers to analyze each aspect of the system to determine what hazards may exist. The behavior of complex systems cannot be inferred or fully understood from individual component behaviors (including events or actions). The system-level behavior must be understood as stemming from the interactions between components within the system as a whole.

Integration of the risk management plan. In order to improve and develop risk management plans integrated into complex problems, risk managers should not focus only on the individual entities but also treat the integrated systems as a whole. Perrow mentioned that "normal accidents" are failures that might cause a trivial impact at the beginning but can spread widely and produce long-lasting effects.[16] These normal accidents occur in multiple integrated complex systems (high risk systems) due to the rich interactions and integration between the parts within the system, as well as between the system in relationship to other systems. This creates a high degree of uncertainty for risk problems.

Robust causal chain of events. Thinking and speaking a holistic language helps risk managers to increase their level of systems thinking to move beyond the simplicity of a cause–effect paradigm, where repeatability is assumed. In complex system problems, it is difficult to trace the causes and their effects, and the existence of repeatability for complex system problem "causes" is problematic. In other words, it is naive to think that we have a linear component of one-problem one-solution in a chaotic environment. It is questionable to believe that managing risks in such an environment using a reductionism (cause–effect) paradigm can achieve the desired overall system safety.[17] The systems-thinking-based paradigm is much more consistent with the environment faced by risk managers.

Less surprises. Emergence presents one of the main obstacles for risk managers. In fact, the hazards that are associated with the emergent behaviors in complex systems especially cannot be determined before the system is put into operation. Such unintended behaviors can easily create countervailing risks that are not necessarily known, or predictable, prior to system deployment and operation. Even though these emergent behaviors are not expected, treating the problem from a holistic perspective will indeed alert risk managers that they need a flexible and resilient risk management plans to cope with any unexpected behaviors or unpredictable events. The high level of uncertainty in complex systems makes it difficult to find one best design that controls and mitigates all risks. Applying a systems thinking perspective, risk managers might need different types of procedures to make sure that they capture accidents that often arise from the interaction and uncertainty among the system components.[18] Complex system has incomplete and shifting boundaries based on the understanding of the complex problem. Keeping this in mind, risk managers need to design risk management plans that are flexible enough to be improved and adjusted to cope with the new events that might occur at any time. As knowledge of a dynamic complex system evolves with corresponding reduction in ambiguities, so too must the response to identify new categories of emergent risk not previously known.

6.8 SUMMARY

Attempting to improve long-term risk management in a complex systems domain by analyzing and changing individual components has often proven to be unsuccessful. The difficulties for risk managers are in coping with the increasing complexities of these systems. Better risk management in complex systems domain requires risk managers to have knowledge from not only the technical elements of the systems but also the inherent human or social, organizational or managerial, and political or policy dimensions. To successfully address complexity, risk managers must shift to a paradigm with a holistic perspective. This paradigm is based on systems thinking and systems theory.

CHAPTER 7

HOW TO FIND SYSTEMS THINKERS

7.1 RISK EVENTS CAUSED BY HUMANS

In Chapter 1, various categories of risk events were described based on causes, origin, and consequences. If we focus on the causes of risk events, then they can be thought of as being caused by either human (manmade) or nonhuman characteristics.[1] Risk events caused by humans stem from a variety of reasons, including poorly designed system, system design parameters that did not anticipate emergent behavior, or even direct inappropriate behaviors and activities of humans. These sources of manmade failures can result from a lack of knowledge of complex systems as well as a lack of systems skills. In fact, many risk events in complex system problems relate more to organizational and individual issues where people are an essential, if not dominant, contributor to the events. These risk events can be classified as having sociotechnical aspects stemming from both technical and social (human) elements as well as the interactions between those elements.[2] There are many examples in which risk events are attributed to human error such as the oil fire in Texas in 2005, the Pacific Garbage Patch in 2008, and the Gulf Oil Spill in 2010—all were related to human (policy, skills, or political) issues. Therefore, holistic systems thinking should be engaged by risk managers to consider the complex nature of failures in manmade systems.

To avoid manmade risk events, risk managers must have the capabilities to deal with new hazards or unpredictable events. As discussed in Chapter 3, there are many risk management tools and techniques that can be used to manage risks. However, there is also a need to have tools and

techniques that can assess the systems skills (systems thinking characteristics) of individuals. In effect, a systemic view, which invokes a more
holistic perspective integral to systems thinking, is necessary for risk managers to deal with complex problems. Without a thorough appreciation of
the dynamic environment and context of complex systems, the conditions
are set for incompatibility between the selected risk analysis approach or
tools, the problem, and the context within which the system is embedded.[3]
This incompatibility between approach, problem, and context most often
produces chaotic and unsustainable risk management plans and leads to
failures in complex problem domains. The ability to recognize this incompatibility is a function of higher-order systems thinking.

7.2 SYSTEMS THINKING FOR RISK MANAGERS

Effectiveness in systems thinking is a critical skill for addressing some of
the most challenging problems of the 21st century, including those faced
by risk managers. Thus, a *systems thinking instrument* is designed to help
individuals who have a propensity for systems thinking when dealing with
complex system problems that cross different domains such as transportation, energy, healthcare, and others.

The systems thinking instrument was developed using a mixed methods approach to collect qualitative and quantitative data to analyze 1,000
different literature sources to derive the systems thinking characteristics that individuals need when addressing complex systems problems.
Following Grounded Theory,[4] a rigorous methodology was executed to
inductively build the framework for systems thinking characteristics.

After deriving the set of systems thinking characteristics (Table 7.3),
an Exploratory and Confirmatory Factor Analysis and Monte Carlo Simu-
lation were used to analyze the dataset obtained from 380 participants. The
results of the analysis showed that the new instrument (web-based survey)
measures and captures the level of systems thinking for individuals.

To check the validity of the systems thinking instrument, multiple
validity checks, including face validity, internal validity, conclusion validity, and content validity, were engaged. The instrument was tested using
different test types to establish the reliability of the instrument, including Cronbach's Alpha Test and all showed excellent results (Table 7.1).
The reliability of the instrument was established, and the validity was
supported by statistical tests.

The outcome of the systems thinking instrument provides a profile
that presents the systems thinking characteristics held by an individual.

Table 7.1. Cronbach's alpha of the systems thinking instrument

Reliability Statistics

Cronbach's Alpha	Cronbach's Alpha Based on Standardized Items	Number of Items
0.83	0.83	39

Table 7.2. Systems thinking instrument score sheet

	a	b		a	b		a	b		a	b		a	b		a	b		a	b
1			7			12			18			24			21			35		
8			2			4			31			25			19			15		
3			20			14			9			37			32			26		
13			10			36			30			27			5			16		
33			11			38			22			28			34			6		
39						17			29			23								

In effect, the instrument develops the degree to which a risk manager's particular systems worldview is compatible with the complexity, uncertainty, ambiguity, and emergency inherent in complex problem environment he must navigate.

The systems thinking instrument consists of 39 binary questions with a score sheet. The score sheet provides the systems thinking profile for an individual (Table 7.2). The outcome of the systems thinking instrument consists of 14 scored scales to measure the 7 main preferences as shown in Table 7.3.

These 14 labels reflect a risk manager's level of systems thinking—thinking that is essential to enhancing risk management within complex systems. As discussed in Chapter 6, managing risks in simple systems (linear systems) is significantly different from managing risks in complex systems (nonlinear systems). In simple systems, the traditional reductionist approach of breaking the systems into manageable elements and then analyzing each element to determine what hazards may exist is sufficient,

Table 7.3. Fourteen scored scales to measure the seven main preferences in systems thinking

Complexity	C	S	Simplicity
Integration	G	A	Autonomy
Interconnectivity	I	N	Isolation
Holism	H	R	Reductionism
Emergence	E	T	Stability
Flexibility	F	D	Rigidity
Embracement over requirements	V	Y	Resistance over requirements

applicable, and useful.[5] However, the interactions and interdependences that exist within and among complex systems make it difficult to analyze each element of the system in isolation, making reductionist approaches tenuous for complex systems.

Risk managers most often operate under the conditions associated with rare events and conditions of *ambiguity* (lack of clarity in understanding), *uncertainty* (incomplete knowledge) *emergence* (unpredictable patterns), and *complexity* (rich dynamic interactions). These conditions are likely to continue as risk managers struggle with the interdisciplinary system problems of the 21st century. Table 7.2 also shows the preferences pairs for the set of systems thinking characteristics. It is important to mention that there are no intrinsically good or bad combinations; it depends solely on the uniqueness of the problem domain within which the risk manager is engaged. However, to deal effectively with the challenges associated with complex systems, risk managers need to be good systems thinkers.

7.3 SEVEN PAIRS OF SYSTEMS THINKING PREFERENCE

As shown in Table 7.3, there are seven pairs of systems thinking preferences. To illustrate, the first pair of preferences deals with the level of complexity, the second with the level of integration, the third with the level of interaction, the fourth with the level of change, the fifth with the level of uncertainty and ambiguity, the sixth with the level of hierarchical view, and the seventh with the level of flexibility. A combination of these preferences determines the level of systems thinking that would indicate a risk manager's predisposition to deal successfully with risk events in complex problem domains from a systems thinking perspective. This does not suggest that risk managers who do not score more to the "systemic" side of thinking on the scales presented cannot be successful in the complex system problem domain. Instead, for more "non-systems thinkers," it suggests that the degree of "systemic nature" of the problem domain might present more difficulty (acknowledgement, frustration, decision, action) due to the probable incompatibility with their worldview. Additionally, the pairs present "bookends" of propensity for an individual to prefer the degree of systemic thinking comfort with their worldview. Worldviews of individuals can, and most certainly will, exist along the spectrum between the "bookends."

Complexity–simplicity (C-S) pair of preferences describes a risk manager's inclination to work in complex problem domains. As mentioned in Chapter 6, one of the main attributes of complex systems is *complexity*. To understand and deal with complexity, safety professionals need to be a complexity-type systems thinker. If a risk manager is on the complexity spectrum (C), he or she probably

- Tends to accept working solutions;
- Enjoys working on problems that have not only technological issues but also the inherent human or social, organizational or managerial, and political or policy dimensions;
- Expects and prepares for unexpected events when managing risks (emergence); and
- Is willing to plan for managing risks in a rapidly changing environment.

If the risk manager is on the simplicity spectrum (S), he or she probably (a) prefers to work on problems that have clear causes, (b) prefers one best solution (one risk management plan) to the problem, and (c) enjoys managing risks on small-scale problems (reduced to direct cause–effect relationships).

Integration–autonomy (G-A) pair of preferences deals with the level of autonomy. Since *integration* is another attribute of complex systems, it is necessary to know the risk manager's comfort zone in dealing with integration. If the risk manager is on the integration spectrum (G), he or she probably

- Understands and appreciates the purpose of the global integration for a system;
- Prefers to make decisions dependent on a multitude of internal and external factors; and

- Focuses more on the overall performance of the systems rather than isolated performance factors.

However, if the risk manager is on the autonomy spectrum (A), then he or she probably (a) leans more toward comfort with independent decisions, (b) focuses more on the individual elements of the systems, and (c) focuses on the local system performance.

Interconnectivity–isolation (I-N) pair of preferences, which pertains to the level of interaction, describes the type of work environment that risk managers prefer: interconnected or isolated. In complex problem domains, rich interaction between the individual systems and their elements presents a considerable challenge for enhancing management of risks. Thus, it is important to have integration type system thinkers to effectively deal with the risks involved from the extensive interactions that are indicative of complex systems. If the risk manager is on the interconnectivity side of the spectrum (I), he or she probably

- Enjoys working on risk management problems within a team;
- Applies a flexible plan to identify and analyze risks;
- Understands the difficulty of having a clear cause–effect paradigm; and
- Focuses on the overall interaction of the system to improve overall management of risks.

On the spectrum of isolation (N), the level of systems thinking for risk manager indicates that he or she (a) enjoys working on problems that have clear cause–effect relationships, (b) leans toward working individually on the problem, and (c) prefers managing risks in small systems with less interaction among the elements.

Emergence–stability (E-T) pair of preferences deals with the level of uncertainty and ambiguity. This level describes a risk manager's preference to make decisions focusing more on emergence or stability. Unintended behaviors and their consequences (emergence) always question risk management strategies in complex systems. As early as 1969, it has been mentioned that all systems are emergent and that there is a need to have emergence type systems thinkers before the selection of risk management tools to identify and analyze risk in complex systems.[6] If the risk manager is on the emergence side of the spectrum (E), he or she probably

- Applies a holistic view in developing and improving risk management plans in complex problems;

- Is comfortable in dealing with uncertainty;
- Prefers working with consideration for nontechnical problems; and
- Follows a general-flexible plan to prepare for any unexpected behaviors.

If the level of systems thinking for risk manager is on the stability spectrum (T), then he or she (a) prepares a detailed plan expecting the details will hold true in the future, (b) avoids working on problems that have a high level of uncertainty and ambiguity, and (c) prefers working on technical problems that are more objective in nature.

Holism–reductionism (H-R) pair of preferences deals with the level of hierarchical view of the system. This level describes a risk manager's predisposition as to how he looks at a problem in complex systems. It is necessary to have a holistic-type systems thinker to deal with the attributes of complex problem domains. If the level of systems thinking for risk manager leans toward the holism side of the spectrum (H), he or she probably

- Focuses more on the whole in solving problems and enhancing management of risks;
- Identifies risks in the system by looking first at the big picture to understand the higher-level interactions; and
- Focuses more on the conceptual ideas instead of following the details inherent in cause–effect solutions.

If the level of systems thinking for risk manager is leaning toward the reductionism side of the spectrum (R), then he or she probably (a) focuses more on the particulars and details in solving problems and (b) prefers to work on problems that have direct cause–effect attributable solutions.

Flexibility–rigidity (F-D) pair of preferences deals with the level of flexibility. This level describes a risk manager's preference in making changes and decisions. A flexible-type systems thinker is necessary to cope with the dynamic nature of complex systems and, therefore, develop a greater degree of robustness in managing risks. If the risk manager is on the flexibility spectrum (F), he or she probably

- Enjoys working on multidimensional problems;
- Reacts to problems as they occur;
- Avoids routine processes; and
- Prepares flexible plans.

The other level of systems thinking is on the rigidity side of the spectrum (D). In this side of the spectrum, risk managers (a) prefer working on well-bounded problems, (b) prepare and work to detailed plans, and (c) enjoy working more by routine.

Embracement over requirements–Resistance over requirements (V-Y) pair of preferences describes a risk manager's inclination to make changes in complex problems. *Embracement over requirements* (from Table 6.1) is another attribute of complex systems where risk managers need to be aware of the rapid changes. An embracement over requirements type systems thinker has a paradigm that is better equipped to manage risks and minimize hazards in complex systems. If the risk manager is on this side of spectrum (V), he or she probably

- Prefers to work in a changing environment where multiple strategies are needed to minimize risks;
- Is apt to take multiple viewpoints into consideration before making a decision regarding risk management strategies;
- Focuses more on the external forces, such as contextual issues, and how such factors can impact risk management in a negative way; and
- Prefers to design a flexible risk management process to cope with the shifting changes in systems' requirements.

The other spectrum of this pair is resistance over requirements (Y). If the risk manager is on this spectrum, then he or she (a) assumes that too many perspectives could create distractions, (b) focuses intensively on the internal forces that directly affect risks of a system, and (c) prefers to work in a stable environment where changes are slow.

As illustrated, these seven main preference pairs determine the level of systems thinking for an individual to deal with complex problems. An effective way to establish an individual framework to deal with complex problems is by understanding the proposed set of systems thinking characteristics. To cope with a highly dynamic environment of complex systems, risk managers need to be more system thinkers (holism) than reductionist thinkers. This is not to criticize the reductionist approach but only to place attention on the necessity of having holistic systems thinkers who can understand the overall picture of the system before identifying and analyzing risks as the level of complexity increases.[7]

By considering this systems thinking instrument, a score may be provided for the individual. This scoring can be facilitated by the list of guiding statement in Table 7.4. For risk managers, this score would

Table 7.4. Seven systems thinking preference pairs

Complexity–Simplicity (C-S)	
Tends to accept working solutions Enjoys working on problems that have human or social, organizational or managerial, and political or policy dimensions Expects the unexpected Designs for rapidly changing environment	Prefers to work on problems with clear causes Prefers one best solution
Integration–Autonomy (G-A)	
Understands the purpose of the global integration Prefers to make decisions dependent on a multitude of internal and external factors Focuses more on the overall performance	Leans more toward comfort with independent decisions Focuses more on the individual elements of the systems Focuses on the local system performance
Interconnectivity–Isolation (I-N)	
Enjoys working on problems with a team Applies a flexible plan Understands the difficulty of having a clear cause–effect paradigm Focuses on the overall interaction of the whole system	Enjoys working on problems that have clear cause–effect relationships Leans toward working individually on the problem Prefers designing for relatively small systems
Emergence–Stability (E-T)	
Applies a holistic view in developing and improving risk management in complex problems Is comfortable dealing with uncertainty Prefers working with consideration for nontechnical problems Follows a general, flexible plan to prepare for any unexpected behaviors	Prepares a detailed plan expecting the details will hold true in the future Avoids working on problems that have a high level of uncertainty and ambiguity Prefers working on technical problems that are more objective in nature

(Continued)

Table 7.4. Seven systems thinking preference pairs (Continued)

Holism–Reductionism (H-R)	
Focuses more on the whole in solving problems Identifies risk in the system by looking first at the big picture Focuses more on the conceptual ideas instead of details	Focuses more on the particulars and details in solving problems Prefers to work on problems that have direct cause–effect attributable solutions
Flexibility–Rigidity (F-D)	
Enjoys working on multidimensional problems Reacts to problems as they occur Avoids routine processes and prepares flexible plans	Prefers working on well-bounded problems Prepares and works to detailed plans Enjoys working more by routine
Embracement over requirements–Resistance over requirements (V-Y)	
Prefers to work in a changing environment Is apt to take multiple viewpoints Focuses more on the external forces Prefers flexible processes to cope with the shifting changes in systems' requirements	Assumes that too many perspectives could create distractions Focuses intensively on the internal forces that directly affect risks to a system Prefers to work in a stable environment where changes are slow

translate to a risk manager's level of systems thinking and his or her inclination to deal with complex problem domains. It is expected that a risk manager might agree with some attributes under each preference within the same pair. For example, some individuals might strongly agree with every attribute related to the "emergence" preference and none of the attributes characteristic of the "stability" preference. These individuals focus more on the whole in solving problems instead of preferring reductionist approaches. Other individuals might agree with some of the "emergence" attributes and at the same time agree with some of the "stability" attributes; this could be quite true and natural. To clarify this, there are no good or bad combinations; there are only variances from one individual to another. However, due to the nature of complex problems, context and environment determine which profiles (combinations) are suitable to engage in developing risk management plans in simple systems or complex systems. Although a different problem domain or system might be presented to a risk manager, the preference in level of thinking does not change with presentation of a different problem or system. Therefore, there is a potential to have incompatibility between an individual's dominant systems thinking worldview and the problem domain or system they are charged to engage.

7.4 SUMMARY

Chapter 1 introduced systems approach and Chapter 4 emphasized the importance of personnel responsible for implementing risk management plans to their success. Since systems thinking is important to understand the complex system problems, a nonspecific domain systems thinking instrument will help risk managers who have the propensity for systems thinking—both for himself and others. The outcome of this instrument is a set of systems thinking characteristics to assist in identification of individuals such as risk managers and members of the risk management team who have the ability to more successfully identify risk and integrate safety in complex systems that require a high level of systems thinking.

The concept underlying the development of the systems thinking instrument and consequently, the set of systems thinking characteristics is based on capturing the systems thinking characteristics that are necessary for individuals (e.g., risk managers) to engage in higher-level (holistic) thinking about complex problems and how they approach these problems. Systems skills are always needed in any domain, especially for risk manager, to effectively integrate safety behaviors in complex problems and reduce risks to a level of insignificance.

NOTES

Chapter 1

1. Dickstein and Flast (2009).
2. Ericson (2005), pp. 14–15.
3. Ericson (2005), pp. 14–15.
4. Modified from the Swiss cheese model by Davidmack, last accessed December 17, 2014 at http://commons.wikimedia.org/wiki/File:Swiss_cheese_model_of_accident_causation.png#filelinks
5. Blunden and Thirlwell (2010), p. 7.
6. Blunden and Thirlwell (2010), p. 8; Dickstein and Flast (2009), p 21.
7. King (2001), p. 7.
8. Global Association of Risk Professionals (2013).
9. Blunden and Thirlwell (2010), p. 9.
10. It is recognized that risk events may originate from outside the system (i.e., its environment); this does not preclude these risk events from analyzes using systems approach.
11. Blunden and Thirlwell (2010), p. 11.
12. King (2001), pp. 25–27.
13. Blunden and Thirlwell (2010), p. 16.
14. Transportation Safety Board of Canada (TSB) (2014).

Chapter 2

1. Haimes (1998).
2. Pinto, McShane, and Bozkurt (2012).
3. Rausand (2005).
4. Rausand (2005).

Chapter 3

1. Rausand (2005).
2. Preliminary Hazard Analysis (2014).
3. Dunjó et al. (2010).
4. Adopted from Lihou (2014).
5. Lihou (2014).
6. Lihou (2014).

7. Titus (2012).
8. UC Berkley (2003).
9. JHA Facilities (2014).
10. JHA Facilities (2014).
11. ASQ (2014).
12. Carlson (2014).
13. Carlson (2014).
14. Adopted from Carlson (2014).
15. Adopted from Carlson (2014).
16. Forrest (2010).
17. Adopted from Greenfield (2000).
18. Greenfield (2000).
19. "Cause and Consequence Analysis" (2014).
20. "Cause-Consequence Analysis" (2014).
21. Adopted from Ericson (2005).
22. "Cause and Consequence Analysis" (2014).
23. HSE (2014).
24. Melchers (2001).
25. Melchers (2001).

Chapter 4

1. Hofstetter et al. (2002).
2. As exemplified in the tools described in Chapter 3, for example, Failure Modes and Effects Analysis, the causal chain of events is established in the process of analyzing risk events in terms of their chances of occurrence and consequences if they occur.
3. The Institutes (2013).

Chapter 5

1. SH&E (2012).
2. Breden (2008).
3. Malivert and Bilal (2015).

Chapter 6

1. Azarnoosh et al. (2006).
2. Senge (1990).
3. Jaradat and Keating (2014); Merrow (2003); Armbruster, Endicott-Popovsky, and Whittington (2013); Katina and Jaradat (2012); Jaradat, Keating, and Bradley (2014).
4. Jablonowski (1995), p. 1.

5. This section is partly derived from works conducted at the Center for System of Systems Engineering at Old Dominion University (2008 to 2014).
6. Harris, Pritchard, and Robins (1995).
7. Mitroff (1998).
8. These attributes were summarized from Leveson (2002), Hitchins (2003), Ackoff (1971), Boardman and Sauser (2006), Cook and Sproles (2000), Crossley (2004), DeLaurentis (2005), and Maier (1998).
9. Leveson (2002); Checkland (1999); Holland (1998); Keating (2009).
10. Sage (1995).
11. Checkland (1993); Hitchins (2003); Jackson (1993); Senge et al. (1994); Hoefler and Mar (1992); Boardman and Sauser (2008).
12. von Bertalanffy (1968).
13. Detailed and complete list of the principles, laws, and concepts can be found in the work of Clemson (1984), Skyttner (2001), Jackson (2003), Adams and Keating (2011), and Adams et al. (2014).
14. Hammitt (2009), p. 2.
15. Mitroff (1998).
16. Perrow (1984).
17. Keating, Padilla, and Adams (2008); Fred (2003); Leveson (2002).
18. Leveson (1995).

Chapter 7

1. Bankoff, Frerks, and Hilhorst (2003); Barton (1969); Korstanje (2011).
2. Jaradat and Keating (2014); Perrow (1984); Jaradat and Pinto (2015).
3. Jaradat (2015).
4. Strauss and Corbin (1990).
5. Azarnoosh et al. (2006); Holloway and Johnson (2006).
6. Simon (1969).
7. Boulding (1956); Clemson (1984).

REFERENCES

Ackoff, R.L. 1971. "Towards a System of Systems Concepts." *Management Science* 17, no. 11, pp. 661–71. doi: http://dx.doi.org/10.1287/mnsc.17.11.661

Adams, K.M., P.T. Hester, J.M. Bradley, T.J. Meyers, and C.B. Keating. 2014. "Systems Theory as the Foundation for Understanding Systems." *Systems Engineering* 17, no. 1, 112–23. doi: http://dx.doi.org/10.1002/sys.21255

Adams, K., and C. Keating. 2011. "Overview of the Systems of Systems Engineering Methodology." *International Journal of System of Systems Engineering* 2, no. 2, pp. 112–19. doi: http://dx.doi.org/10.1504/ijsse.2011.040549

Armbruster, G., B. Endicott-Popovsky, and J. Whittington. 2013. "Threats to Municipal Information Systems Posed by Aging Infrastructure." *International Journal of Critical Infrastructure Protection* 6, no. 3, pp. 123–31. doi: http://dx.doi.org/10.1016/j.ijcip.2013.08.001

ASQ. 2014. "Failure Mode Effects Analysis (FMEA)." http://asq.org/learn-about-quality/process-analysis-tools/overview/fmea.html (accessed October 29).

Azarnoosh, H., B. Horan, P. Sridhar, A.M. Madni, and M. Jamshidi. July 24–26, 2006. "Towards Optimization of a Real-World Robotic-Sensor System of Systems." *Proceedings of World Automation Congress (WAC)*. Budapest, Hungary: IEEE.

Bankoff, G., G. Frerks, and D. Hilhorst, eds. 2003. *Mapping Vulnerability: Disasters, Development and People*. Sterling, VA: Earthscan. ISBN 1-85383-964-7.

Barton, A.H. 1969. *Communities in Disaster: A Sociological Analysis of Collective Stress Situations*. Garden City, NY: Doubleday.

Blunden, T., and J. Thirlwell. 2010. *Mastering Operational Risk: A Practical Guide to Understanding Operational Risk and How to Manage it*. Harlow, England: Financial Times Prentice Hall.

Boardman, J., and B. Sauser. April 24–26, 2006. "System of Systems - the Meaning of of." *IEEE International Conference on System of Systems Engineering*. Los Angeles, CA: IEEE.

Boardman, J., and B. Sauser. 2008. *Systems Thinking: Coping with 21st Century Problems*. Boca Raton, FL: CRC Press.

Boulding, K.E. 1956. "General Systems Theory-the Skeleton of Science." *Management Science* 2, no. 3, pp. 197–208. doi: http://dx.doi.org/10.1287/mnsc.2.3.197

Breden, D. 2008. "Monitoring the Operational Risk Environment Effectively." *Journal of Risk Management in Financial Institutions* 1, no. 2, pp. 156–64. doi: http://www.ngentaconnect.com/content/hsp/jrmfi/2008/00000001/00000002/art00004

Carlson, C. 2014. "Understanding and Applying the Fundamentals of FMEAs." Paper presented at the 2014 IEEE Reliability and Maintainability Symposium, Tucson, AZ.

"Cause and Consequence Analysis." 2014. Safety to Safety http://www.safety-s2s. eu/modules.php?name=s2s_wp4&idpart=4&idp=54 (accessed September 7).

"Cause-Consequence Analysis." 2014. Ramentor. http://www.ramentor.com/ theory/cause-consequence-analysis/ (accessed September 7).

Checkland, P. 1993. *Systems Thinking, Systems Practice*. New York: John Wiley & Sons.

Checkland, P. 1999. *Systems Thinking, Systems Practice*, 2nd ed. New York: John Wiley & Sons.

Clemson, B. 1984. *Cybernetics: A New Management Tool*. Tunbridge Wells, Kent: Abacus Press.

Cook, S.C., and N. Sproles. 2000. "Synoptic Views of Defense Systems Development." *Proceedings of SETE SESA and ITEA*, Brisbane, Australia.

Crossley, W.A. 2004. "System of Systems: An Introduction of Purdue University Schools of Engineering's Signature Area." Paper presented at the Engineering Systems Symposium, Cambridge, MA.

DeLaurentis, D. 2005. "Understanding Transportation as a System-of-Systems Design Problem." *Proceeding of the 43rd AIAA Aerospace Sciences Meeting and Exhibit*, Vol. 1, Blacksburg, VA.

Dickstein, D.I., and R.H. Flast. 2009. *No Excuses: A Business Process Approach to Managing Operational*. Hoboken, NJ: Wiley-Interscience.

Dunjó, J., V. Fthenakis, J.A. Vílchez, and J. Arnaldos. 2010. "Hazard and Operability (HAZOP) Analysis. A Literature Review." *Journal of Hazardous Materials* 173, no. 1–3, pp. 19–32. doi: http://dx.doi.org/10.1016/j.jhazmat. 2009.08.076

Ericson, C.A. 2005. *Hazard Analysis Techniques for System Safety*. Hoboken, NJ: Wiley-Interscience.

Forrest, G. 2010. "Quick Guide to Failure Mode and Effects Analysis." Isixsigma. http://www.isixsigma.com/tools-templates/fmea/quick-guide-failure-mode-and-effects-analysis/ (accessed May 24, 2014).

Fred, A.M. 2003. *On the Practice of Safety*. 3rd ed. Hoboken, NY: John Wiley & Sons.

Global Association of Risk Professionals. 2013. *Operational Risk Management*. http://www.garp.org/media/673303/operational%20risk%20slides.pdf (accessed September 29).

Greenfield, M.A. 2000. "Risk Management Tools." Presented at NASA Langley Research Center, Hampton, VA.

Haimes, Y.Y. 1998. "Risk Modeling, Assessment, and Management." New York: John Wiley & Sons.

Hammitt, K. 2009. "Characterizing Social Preferences for Health and Environmental Risks." Paper presented at the Decision Analysis: Supporting Environmental Decision Makers Workshop, The Office of Research and

Development's National Risk Management Research Laboratory, Harvard University, Boston.

Harris, C., M. Pritchard, and M. Robins. 1995. *Engineering Ethics: Concepts and Cases*. Belmont, CA: Wadsworth Publishing Co.

Hitchins, D. 2003. *Advanced Systems Thinking, Engineering, and Management*. Norwood, NJ: Artech House.

Hoefler, B.G., and B.W. Mar. 1992. "Systems Engineering Methodology for Engineering Planning Applications." *The Journal of Professional Issues in Engineering Education and Practice* 118, no. 2, pp. 113–28. doi: http://dx. doi.org/10.1061/(asce)1052-3928(1992)118:2(113)

Hofstetter, P., J.C. Bare, J.K. Hammitt, P.A. Murphy, and G.E. Rice. 2002. "Tools for Comparative Analysis of Alternatives: Competing or Complementary Perspectives?" *Risk Analysis* 22, no. 5, pp. 833–51. doi: http://dx.doi. org/10.1111/1539-6924.00255

Holland, J. 1998. *Emergence*. New York: Plenum.

Holloway, C.M., and C.W. Johnson. 2006. "Why System Safety Professionals Should Read Accident Reports." In *The First IET International Conference on System Safety*, ed. T. Kelly. Savoy Place, London: Institute of Engineering and Technology.

HSE. 2014. "ALARP 'at a glance.'" http://www.hse.gov.uk/risk/theory/ alarpglance.htm (accessed February 6).

Jablonowski, M. 1995. "Recognizing Knowledge Imperfection in the Risk Management Process." *Proceedings of ISUMA-NAFIPS '95 the Third International Symposium on Uncertainty Modeling and Analysis and Annual Conference of the North American Fuzzy Information Processing Society*, pp. 1–4. College Park, MD: IEEE.

Jackson, M. 1993. "The System of Systems Methodologies: A Guide to Researchers." *The Journal of the Operational Research Society* 44, no. 2, pp. 208–09. doi: http://dx.doi.org/10.1057/jors.1993.42

Jackson, M. 2003. *Systems Thinking: Creative Holism for Managers*. Chichester, UK: John Wiley & Sons.

Jaradat, R.M. 2015. "Complex System Governance Requires Systems Thinking – How to Find Systems Thinkers." *International Journal of System of Systems Engineering* 6, no. 1–2, pp. 53–70. doi: http://dx.doi.org/10.1504/ ijsse.2015.068813

Jaradat, R., and C. Keating. 2014. "Fragility of Oil as a Complex Critical Infrastructure Problem." *International Journal of Critical Infrastructures Protection* 7, no. 2, pp. 86–99. doi: http://dx.doi.org/10.1016/j.ijcip.2014.04.005

Jaradat, R., C. Keating, and J. Bradley. 2014. "A Histogram Analysis for System of Systems." *International Journal of System of Systems Engineering* 5, no. 3, pp. 193–227. doi: http://www.inderscience.com/info/inarticle.php?artid=65750

Jaradat, R.M., and A.C. Pinto. 2014. "Development of a Framework to Evaluate Human Risk in a Complex Problem Domain." *International Journal of Critical Infrastructures* 11, no. 2, pp. 148–66. doi: http://dx.doi.org/10.1504/ ijcis.2015.068614

JHA Facilities. 2014. "Welding Pipes Using a Torch." UDEL http://www.facilities. udel.edu/docs/JHA/PlumbingShop/PP11WeldPipesUsingTorch.pdf (accessed May 8).

Katina, P., and R. Jaradat. 2012. "A Three-Phase Framework for Elicitation of Infrastructure Requirements." *International Journal of Critical Infrastructures* 8, no. 2/3, pp. 121–33. doi: http://dx.doi.org/10.1504/ijcis.2012.049032

Keating, C.B. 2009. "Emergence in System of Systems." In *System of Systems Engineering: Innovations for the 21st Century*, ed. M. Jamshidi, 169–90. Hoboken, NJ: John Wiley & Sons.

Keating, C., J. Padilla, and K. Adams. 2008. "System of Systems Engineering Requirements: Challenges and Guidelines." *Engineering Management Journal* 20, no. 4, pp. 24–31. doi: http://dx.doi.org/10.1080/10429247.2008 .11431785

King, J.L. 2001. *Operational Risk : Measurement and Modeling*. New York: Wiley.

Korstanje, M. 2011. "Swine Flu, Beyond the Principle of Resilience." *International Journal of Disaster Resilience in the Built Environment* 2, no. 1, pp. 59–73. doi: http://dx.doi.org/10.1108/17595901111108371

Leveson, N.G. 1995. "Safety as a System Property." *Communications of the ACM* 38, no. 11, 146–60. doi: http://dx.doi.org/10.1145/219717.219816

Leveson, N.G. 2002. *System Safety Engineering: Back to the Future*. Cambridge, MA: MIT Press.

Lihou, M. 2014. "Hazard & Operability Studies." Hazop Manager Version 7.0. http://www.lihoutech.com/hazop1.htm (accessed July 8).

Maier, M.W. 1998. "Architecting Principles for Systems-of-Systems." *Systems Engineering* 1, no. 4, pp. 267–84. doi: http://dx.doi.org/10.1002/(sici)1520-6858(1998)1:4<267::aid-sys3>3.0.co;2-d

Malivert, A., and A. Bilal. 2015. "Risk Database Structure Project 4." SWEngGrowth Wikispaces. http://swenggrowth.wikispaces.com/x-5.0%20 Mechanism-5.3%20Risk%20Database%20Structure (accessed February 15).

Melchers, R.E. 2001. "On the ALARP Approach to Risk Management." *Reliability Engineering & System Safety* 71, no. 2, pp. 201–08. doi: http://dx.doi. org/10.1016/s0951-8320(00)00096-x

Merrow, E.W. 2003. "Mega-Field Developments Require Special Tactics, Risk Management." *Management & Economics*. http://projectcontrolsonline.com/ portals/0/primvera-com-au/ipa_megafield_developments.pdf

Mitroff, I. 1998. *Smart Thinking for Crazy Times*. San Francisco, CA: Berrett-Koehler Publisher Inc.

Perrow, C. 1984. *Normal Accidents: Living with High-Risk Technologies*. New York: Basic Books.

Pinto, C.A., M.K. McShane, and I. Bozkurt. 2012. "System of Systems Perspective on Risk: Towards a Unified Concept." *International Journal of System of Systems Engineering* 3, no. 1, pp.33–46. doi: http://dx.doi.org/10.1504/ ijsse.2012.046558

"Preliminary Hazard Analysis." 2014. A Gateway for Plant and Process Safety. http://www.safety-s2s.eu/modules.php?name=s2s_wp4&idpart=4&idp=50 (accessed August 11).

Rausand, M. 2005. "Preliminary Hazard Analysis." Norwegian University of Science and Technology, Trondheim, Norway. http://frigg.ivt.ntnu.no/ross/slides/pha.pdf (accessed May 1, 2014).

Sage, P. 1995. "Risk Management Systems Engineering Systems, Man and Cybernetics, Intelligent Systems for the 21st Century." Paper presented at the IEEE International Conference, Vancouver.

Senge, P. 1990. *The Fifth Discipline: The Art and Practice of the Learning Organization.* New York: Doubleday.

Senge, P.M., C. Roberts, R.B. Ross, B.J. Smith, and A. Kleinter. 1994. *The Fifth Discipline Fieldbook: Strategies and Tools for Building a Learning Organization.* New York: Doubleday.

SH&E. 2012. "Monitoring Control Measures." https://nationalvetcontent. edu.au/alfresco/d/d/workspace/SpacesStore/fa878b4d-7172-4094-891f-a350bf45e603/14_04/toolbox14_04/unit2_she/section2_risk_assessment/ lesson2_control_measures.htm (accessed December 25, 2014).

Simon, H. 1969. *The Science of the Artificial.* Cambridge, MA: MIT Press.

Skyttner, L. 2001. *General Systems Theory: Ideas & Applications.* River Edge, NJ: World Scientific.

Strauss, A., and J. Corbin 1990. *Basics of Qualitative Research: Grounded Theory Procedures and Techniques.* Newbury Park, CA: Sage.

The Institutes. 2013. "Risk Treatment." http://www.theinstitutes.org/comet/programs/arm/assets/arm55-chapter.pdf (accessed November 29).

Titus, J.B. 2012. "Machine Safety: What Differentiates a Hazard Analysis from a Risk Assessment?" Control Engineering. http://www. controleng.com/single-article/machine-safety-what-differentiates-a-hazard-analysis-from-a-risk-assessment/914879af196c862ba7dfc14f020b56e6. html (accessed October 15, 2013).

Transportation Safety Board of Canada. 2014. Lac-Mégantic Runaway Train and Derailment Investigation Summary. Gatineau QC. http://www.tsb.gc.ca/eng/rapports-reports/rail/2013/r13d0054/r13d0054-r-es.pdf

UC Berkley. 2003. *Job Safety Analysis, EH&S Fact Sheet.* Berkley, CA: University of California Press.

von Bertalanffy, L. 1968. *General Systems Theory.* New York: Brazillier.

INDEX

OTHER TITLES IN OUR INDUSTRIAL AND SYSTEMS ENGINEERING COLLECTION

William R. Peterson, Editor

Idea Engineering: Creative Thinking and Innovation
by La Verne Abe Harris

Cultural Influences in Engineering Projects
by Morgan E. Henrie

FORTHCOMING TITLES FOR THIS COLLECTION

Creating and Deploying Successful Surveys
by Rafael Ernesto Landaeta

Lean Six Sigma Made S.I.M.P.L.E.: An Integrated Continuous Improvement Framework for Current and Future Leaders
by Ertunga C. Ozelkan and Ian Cato

Applying Inventory Optimization
by Daniel Zavala and Diana Salazar

Momentum Press is one of the leading book publishers in the field of engineering, mathematics, health, and applied sciences. Momentum Press offers over 30 collections, including Aerospace, Biomedical, Civil, Environmental, Nanomaterials, Geotechnical, and many others.

Momentum Press is actively seeking collection editors as well as authors. For more information about becoming an MP author or collection editor, please visit http://www.momentumpress.net/contact

---

Announcing Digital Content Crafted by Librarians

Momentum Press offers digital content as authoritative treatments of advanced engineering topics by leaders in their field. Hosted on ebrary, MP provides practitioners, researchers, faculty, and students in engineering, science, and industry with innovative electronic content in sensors and controls engineering, advanced energy engineering, manufacturing, and materials science.

Momentum Press offers library-friendly terms:

- perpetual access for a one-time fee
- no subscriptions or access fees required
- unlimited concurrent usage permitted
- downloadable PDFs provided
- free MARC records included
- free trials

The **Momentum Press** digital library is very affordable, with no obligation to buy in future years.

For more information, please visit **www.momentumpress.net/library** or to set up a trial in the US, please contact **mpsales@globalepress.com.**

www.ingramcontent.com/pod-product-compliance
Lightning Source LLC
Chambersburg PA
CBHW052014230326
41598CB00078B/3412